Edition KWV

Die „Edition KWV" beinhaltet hochwertige Werke aus dem Bereich der Wirtschaftswissenschaften. Alle Werke in der Reihe erschienen ursprünglich im Kölner Wissenschaftsverlag, dessen Programm Springer Gabler 2018 übernommen hat.

Weitere Bände in der Reihe http://www.springer.com/series/16033

Michael Hoeck

Cooperation and Technological Endowment in International Joint Ventures: German Firms in China

 Springer Gabler

Michael Hoeck
Proyecta Ventures GmbH
Herrsching a. Ammersee, Germany

Bis 2018 erschien der Titel im Kölner Wissenschaftsverlag, Köln
Dissertation D 82 (Diss. RWTH Aachen, 2008)

Edition KWV
ISBN 978-3-658-24354-8   ISBN 978-3-658-24355-5   (eBook)
https://doi.org/10.1007/978-3-658-24355-5

Library of Congress Control Number: 2019931815

Springer Gabler
© Springer Fachmedien Wiesbaden GmbH, part of Springer Nature 2008, Reprint 2019
Originally published by Kölner Wissenschaftsverlag, Köln, 2008

This Springer Gabler imprint is published by the registered company Springer Fachmedien Wiesbaden GmbH
part of Springer Nature
The registered company address is: Abraham-Lincoln-Str. 46, 65189 Wiesbaden, Germany

Foreword

High growth rates and the absolute size of the Chinese market have contributed to a dramatic rise in foreign direct investments by German firms in China. However, the investment boom has been accompanied by critical reports that point out the difficulty of protecting intellectual property in China and that German as well as other foreign investors regularly face local diffusion of technical knowledge.

The resulting fear of knowledge drain influences the market entry behaviour of (potential) German investors into China: They either refrain from technologically sophisticated investments or only invest after careful consideration of various influence factors. The dilemma that German firms face relating to market entry to China has been expressed by Heinrich v. Pierer, the former CEO of Siemens AG by the words "The risk of not being present in China is larger than the risk of being present"

A frequent choice for market entry to China is the international joint venture with a Chinese partner. While the use of international joint ventures has been the only legally allowed market entry form in many industries until well into the 90s, today investors are free to choose a market entry form to a large extent. With the exception of strategically significant industries, foreign investors can found 100% foreign-owned subsidiaries on Chinese soil.

International joint ventures are regarded as adequate market entry forms if complex technological knowledge is to be transferred to the new location. However, international joint ventures also represent an easy way for Chinese firms to absorb technological knowledge without authorization and to either use it themselves or to provide access to it to third parties.

Michael Hoeck investigates the character and the degree of technology transfer into international joint ventures, using the example of German industrial firms in China.

The two central research questions that are investigated by the author are „What factors influence the sophistication of the technological endowment that an international joint venture in China receives from its German parents?" and „In what way do strategic considerations regarding inter-firm cooperation and knowledge sharing influence the foreign investor's technology transfer be-haviour?". His work presents a sophisticated and detailed analysis of the topic and presents novel insights to both researchers and practitioners.

Mannheim, September 2008   Prof. Dr. Michael Woywode

Preface

The topics 'international technology transfer' and 'costs of benefits of knowledge sharing between firms' have widely been discussed in both theoretical literature as well as management press. The abundance of literature (well over 10.000 research articles) that has been created – with accelerating speed since the end of the 70s - has first discouraged me from trying to find new insights. It also struck me that history seems to repeat itself – much literature in the 70s describes the worries of U.S. firms when investing in the former low cost country Germany.

However, the topic is today as relevant as ever and as far as I can judge, several factors make the recent Chinese case a special one. It is the speed of economic developments, the size of the Chinese economy, and the fact that decisions by foreign investors are strongly guided by a foreign investment framework designed by the Chinese government to absorb as much technology as possible the shortest possible time.

From a theoretical perspective, recently developed or improved frameworks of cooperation and knowledge exchange have not been tested empirically to a sufficient degree, so the case of German industrial firms in China provides a good opportunity to gather additional evidence.

I therefore hope that my analysis on the behaviour of German investor firms with respect to the sophistication of technology deployed to international joint ventures in China and the degree of insight granted to local Chinese partners finds interested readers – both researchers and practitioners.

The time spent as a research assistant at the Economics Faculty of RWTH Aachen University under the supervision of Prof. Dr. Michael Woywode has contributed greatly to my formation as both a researcher and as a person. I therefore thank above all Michael Woywode for his various forms of guidance and the freedom he provided me with. I also thank my former colleagues Dr. Garnet Kasperk, Dr. Markus Helfen, Achim Oberg and Dr. Matthias Hansch as well as my "extended set" of supervisors, namely Prof. Dr. Oliver Lorz, Prof. Dr. Eberhard Feess, and PD Dr. Markus Reihlen.

This work is dedicated to both my parents as well as my wife. My parents have supported me from day one and I am fortunate to still benefit from their guidance and advice today. Finding and marrying my wife Patricia is probably the most important outcome of my PhD time, and she has fulfilled me with happiness and great optimism for all developments to come. She also has been my most important sparring partner and is probably the person who dedicated the most time and effort to support my work. Finally, I greet the crowd of Nuffield College at Oxford University, who welcomed me with great hospitality during the time I spent there when finishing my thesis.

Munich, September 2008   Michael Hoeck

Table of Contents

List of Tables XIII

List of Figures XV

List of Abbreviations XVII

Chapter I: Introduction 1

1. Description of the Research Topic 1

2. Literature Review and Research Question 5

3. Structure of the Thesis 7

Chapter II: Literature Review 11

1. On Technology.. 11

    1.1. Technology: Definitions and Characteristics 11

        1.1.1. Definitions.. 11

        1.1.2. Characteristics of Technology... 13

        1.1.3. Technology Types and Industry Classifications 14

    1.2. Indicators for Technological Sophistication 17

        1.2.1. Which Functional Activities are Carried Out?............................. 17

        1.2.2. The Sophistication of an Organisation's Technological Resource Endowment 19

        1.2.3. The Sophistication of an Organisation's Technological Capabilities 21

        1.2.4. The Performance of an Organisation as a Proxy for its Technological Skills ... 23

    1.3. International Technology Transfer (ITT).. 24

        1.3.1. ITT in the Context of this Study... 25

        1.3.2. Relevant Issues in ITT Management... 28

2. Technological Leadership and International Market Entry 31

    2.1. The Role of Technological Leadership for International Market Entry.... 32

        2.1.1. Explanations for International Market Entry 32

        2.1.2. The Resource-Based View 35

        2.1.3. Technological Resources and International Market Entry 37

    2.2. The Choice for a Market Entry Form and Technology Transfer 38

        2.2.1. Overview of Market Entry Forms 39

        2.2.2. The Choice for a Market Entry Form... 41

        2.2.3. Technology Transfer and the Choice for an IJV 45

3. The International Joint Venture as a Channel for Technology Transfer 49

    3.1. The (International) Joint Venture.. 49

        3.1.1. Definitions... 49

        3.1.2. Strategic Objectives for International Joint Ventures Formation....... 51

        3.1.3. IJV Classifications and Allocation of Partner Roles......................... 53

3.2. Technology Transfer within International Joint Ventures 56

4. China as a Target Market for International Market Entry 61

4.1. The Political and Economic Development in China since 1978 62

4.2. Chinese Policies and Regulations regarding Technology Transfer 66

4.2.1. General Policy of the Chinese Government regarding Foreign Technology 66

4.2.2. Regulatory Framework and the Government's Role.. 68

4.2.3. Knowledge Drain and the IP Protection Regime in China................................. 71

4.3. Legal Provisions for Relevant Market Entry Forms 77

4.3.1. Representative Office and Licensing 77

4.3.2. Foreign Invested Enterprises... 78

4.3. Technology Transfer and FDI to China from the World and Germany 83

4.3.1. Technology Transfer to China.. 83

4.3.2. FDI from Germany to China 87

5. Recapitulation 89

Chapter III: Theoretical Considerations and Derivation of Hypotheses 91

1. Cooperation in IJVs and Technology Transfer from a
Theoretical Perspective 91

1.1. A Simple Prisoner's Dilemma Paradigm and its Application to
Knowledge Exchange.. 95

1.2. A Prisoner's Dilemma for Technology Transfer to IJVs 99

1.2.1. Variable Names and Descriptions 100

1.2.2. Basic Insights 105

1.2.3. Incorporating a Local Firm's Cooperation Probability -
the Relevance of Trust.. 108

1.2.4. Taking Repeated Interaction and Reciprocity into Account 112

2. Discussion of General Influence Factors 119

2.1. External Influence Factors 121

2.1.1. Local Market Conditions.. 121

2.1.2. Industry-Specific Market Attractiveness.. 125

2.1.3. Differences in Cost Levels 126

2.1.4. Government Incentives 127

2.2. Internal Influence Factors... 128

2.2.1. Characteristics of the Technology Sending Firm.. 129

2.2.2. Product and Technology Characteristics.. 132

2.2.3. The Technology Recipient and Practical Implementation Barriers 135

Chapter IV: Empirical Study...**139**

1. Research Design...139

  1.1. Choice of the Empirical Setting and Structure of the Investigation.......................140

  1.2. Measurement Concepts for the Dependent Variable...144

    1.2.1. Overview of Measuring Concepts...145

    1.2.2. Measuring T(JV) 147

    1.2.3. Measuring T(JV)/T(HQ) 152

    1.2.4. Measuring Shared Technological Insight..156

    1.2.4. Possible Measures of Technology Transfer that are Not Pursued 159

2. Empirical Research Results 160

  2.1. Course of the Investigation and Sample Description 161

  2.2. Descriptive Results..167

    2.2.1. Characteristics of Joint Ventures and the Cooperative Setting 168

    2.2.2. Technological Sophistication of Joint Ventures..171

    2.2.3. Technology Endowment of Joint Ventures...175

  2.3. Statistical Investigation of Hypotheses 180

    2.3.1. Findings related to Technoware 182

    2.3.2. Findings related to Inforware 187

    2.3.3. Findings related to Humanware 192

    2.3.4. Findings related to Orgaware 196

    2.3.5. Findings related to Capabilities..200

3. Summary and Discussion of Results 204

  3.1. Summary of Results and Evaluation of Hypotheses 204

  3.2. Discussion of Results 210

Chapter V: Conclusion **215**

1. Research Limitations 215

  1.1. Sample Characteristics 215

  1.2. Inference of Technology Transfer Commitment..216

  1.3. Ambiguity of Causal Chains 217

2. Conclusion and Suggestions for further Research......................219

VI. References and Appendices **225**

1. Reference List...225

2. Important Chinese Laws and Regulations 259

3. Appendices...261

  Appendix 1: Technology Fields 261

  Appendix 2: Levels of Sophistication of Technological Resources 262

  Appendix 3: Description of Expert Interviews 263

  Appendix 4: Overview of Variables that Correspond to Hypotheses 264

Appendix 5: Overview of General Influence Factors and Corresponding Variables 265
Appendix 6: Qualitative Descriptions of Joint Ventures 267
Appendix 7: Survey Questionnaire (English Version)... 269

List of Tables

Table 1: System for Classifying Technologies 16

Table 2: Levels of Sophistication of the Components of Technology 20

Table 3: Market and Non-Market Channels of ITT between Firms 27

Table 4: Theories for the Explanation of International Market Entry...................................... 33

Table 5: Classification of Market Entry Forms... 39

Table 6: Empirical Findings on Technology and the MEF Choice (Focus on IJVs).............. 47

Table 7: Types of International Joint Ventures.. 54

Table 8: Advantages and Disadvantages of Joint Ventures 55

Table 9: Laws and Regulations regarding IP Protection in China 75

Table 10: Variable Names for Model.. 100

Table 11: List of Hypotheses 119

Table 12: Scores of Germany and China in Hofstede's Cultural Dimensions...................... 122

Table 13: Influence of Investment Regulations on Market Entry Form in China.................. 141

Table 14: Information on Individual Dimensions of Technological Resources 149

Table 15: Information on Individual Dimensions of Technological Capabilities.................. 150

Table 16: All Correlations for Variables Expressing T(JV) 151

Table 17: T(JV)/ T(HQ) for Technological Resources as Dependent Variable.................... 153

Table 18: T(JV)/T(HQ) for Technological Capabilities as Dependent Variable 154

Table 19: All Correlations for Variables Expressing T(JV)/T(HQ) 155

Table 20: Descriptive Statistics for *PartnerInsight*.. 156

Table 21: Descriptive Statistics for Indicators Regarding Shared Technological Insight 158

Table 22: Background of Joint Ventures – Preferred Market Entry Choice? 171

Table 23: Correlations of Measures for *T(JV)* and *T(JV)/T(HQ)* to *PartnerInsight* 179

Table 24: Cases in Sample for which the Sophistication Level of P exceeds HQ or JV 181

Table 25: Regression Results for *TechnowareJV* 183

Table 26: Regression Results for *TechnowareRel* 185

Table 27: Regression Results for *Shared_Technoware* 186

Table 28: Regression Results for *InforwareJV* 188

Table 29: Regression Results for *InforwareRel* 190

Table 30: Regression Results for *Shared_Inforware* 191

Table 31: Regression Results for *HumanwareJV* 193

Table 32: Regression Results for *HumanwareRel* 195

Table 33: Regression Results for *Shared_Humanware* 196

Table 34: Regression Results for *OrgawareJV*... 197

Table 35: Regression Results for *OrgawareRel*.. 199

Table 36: Regression Results for *Shared_Orgaware*.. 200

Table 37: Regression Results for *CapabilitiesJV*.. 201

Table 38: Regression Results for *CapabilitiesRel*... 202

Table 39: Regression Results for *Shared_Capabilities*.. 203
Table 40: Overview of Statistical Results for Evaluation of Hypotheses 208

List of Figures

Figure 1: Germany's World Share for Selected Innovation Indicators (in %).......................... 2
Figure 2: Top 10 Sources of FDI to China in 2005 (in USD million) 3
Figure 3: Structure of the Thesis 8
Figure 4: Technology and the Production Process.. 12
Figure 5: Value Chain Model.. 18
Figure 6: A Basic Model of the Technology Transfer Process 29
Figure 7: The Resource-Based View of the Firm 36
Figure 8: Market Entry Modes and Capital and Management Commitment...................... 41
Figure 9: Summary of Factors Influencing the Entry Form Decision................................ 42
Figure 10: Internal vs. External Transfer Channels for International Market Entry 43
Figure 11: The Eclectic Paradigm and a Firm's Market Entry Choice 44
Figure 12: Goals of Chinese and Foreign Partners in International Alliances...................... 60
Figure 13: Transition of the Chinese Economic System since 1978.................................... 62
Figure 14: China's Gross Domestic Product between 1978 and 2005.................................. 64
Figure 15: Chinese Imports and Exports between 1978 and 2005....................................... 65
Figure 16: Technology Flows to China between 1978 and 2005.. 84
Figure 17: Foreign Direct Investments by Entry Form between 1979 and 2001 85
Figure 18: The Value of Foreign Direct Investment to China 1998-2005 86
Figure 19: Top 10 Sources of FDI to China in 2005 (in USD million) 87
Figure 20: German Stock of FDI in China and Number of German Firms in China.............. 88
Figure 21: Representation of the Basic Prisoner's Dilemma 96
Figure 22: Prisoner's Dilemma Framework of Knowledge Transfer 98
Figure 23: Prisoner's Dilemma Framework for German-Chinese IJVs................................ 101
Figure 24: Computation of Critical Values for Knowledge Transfer 107
Figure 25: Repeated Games 113
Figure 26: Participating Firms - Number of Employees... 162
Figure 27: Background of Respondents - Positions 163
Figure 28: Background of Joint Ventures - Turnover... 164
Figure 29: Background of Joint Ventures – Age of JV and Ownership Share 165
Figure 30: Background of Joint Ventures – Capital Investments and Contract Duration....... 166
Figure 31: Background of Joint Ventures – Joint Venture Types.. 167
Figure 32: Functional Activities Carried Out by Joint Ventures 168
Figure 33: Relative Responsibility for Functional Activities in Joint Ventures 169
Figure 34: Results regarding the Nature of Interaction between Partners 170
Figure 35: Production Process Types... 171
Figure 36: How Innovative and Complex is the Technology Employed? 172
Figure 37: Share of Value Creation by Joint Ventures 173
Figure 38: Number of Products Produced by Life Cycle Stage in Europe and China........... 174

Figure 39: Average Levels of Technological Resources for HQ, JV and the Partner 175
Figure 40: Average Levels of Technological Capabilities for HQ, JV and the Partner......... 177
Figure 41: Degree of Technological Insight of the JV Partner 178
Figure 42: *PartnerInsight* and *HumanwareRel*... 179

List of Abbreviations

ADB	Asian Development Bank
AG	Aktiengesellschaft
ASEAN	Association of Southeast Asian Nations
B2B	Business to Business
bn	Billion
BoD	Board of Directors
BOT	Build-Operate-Transfer
CEO	Chief Executive Officer
CJV	Contractual Joint Venture
Co.	Corporation
CPC	Communist Party of China
DC	Developed Country
DSU	Dispute Settlement Understanding
EJV	Equity Joint Venture
GATS	General Agreement on Trade in Services
GATT	General Agreement on Tariffs and Trade
GNP	Gross National Product
IC	Intellectual Capital
IJV	International Joint Venture
IMF	International Monetary Fund
IP	Intellectual Property
IT	Information Technology
ITT	International Technology Transfer
JV	Joint Venture
LDC	Least Developed Countries
MEF	Market Entry Form
MNC	Multinational Corporation
MNE	Multinational Enterprise
MOFCOM	Ministry of Commerce of the People's Republic of China
MOFERT	Ministry of Foreign Economic Relations and Trade of the People's Republic of China
NBS	National Bureau of Statistics of China
R&D	Research & Development
TNC	Transnational Corporation
TRIPS	Agreement on Trade-Related Aspects of Intellectual Property Rights
TRPM	Trade Policy Review Mechanism
USSR	Union of Soviet Socialist Republics
WFOE	Wholly Foreign-Owned Enterprise

WTO   World Trade Organisation

Chapter I: Introduction

1. Description of the Research Topic

The risk of not being present in China is larger than the risk of being present.
Heinrich v. Pierer as CEO of Siemens AG in 2004 [1]

When firms approach a new foreign target market, they decide on the logic of value creation of their foreign activities, the scope of their activities in the target market, and the governance structure for these activities (Gulati and Singh 1998). Market entry should enable the foreign investor to deploy, maintain, and develop the assets and capabilities that allow the local operations to realise the logic of value creation (Kogut and Zander 1993, Root 1994).

However, the decision to endow foreign operations with certain assets – especially intangible assets such as technology – can have strong repercussions on the firm's future development if IP protection in the target country is weak and the asset deployment causes an increase in imitation risk (Kabiraj and Marjit 1993, Lin and Saggi 1999). Diffusion of technology can then undermine the competitive position of the foreign investor (Reich and Mankin 1986, Bennett et al. 2001, Goh 2004, Feess et al. 2006).

This thesis investigates a setting in which these potential repercussions are evident and widely cited: The market entry of German industrial firms to China by means of international joint ventures (IJVs) and the associated transfer of technology by German investors to the joint venture in the target market.

German firms rely heavily on their technological performance when competing in international markets. Quoting the German Federal Ministry of Education and Research:

> *"Germany's technological performance is essential for German companies' success in international technological competition. [...] Technological performance is documented by new, innovative products and processes which can compete on international markets"* (BMBF 2006).

---

[1]   Original German text: „Das Risiko in China nicht dabei zu sein, ist größer als das, dabei zu sein". Source: von Pierer (2004).

© Springer Fachmedien Wiesbaden GmbH, part of Springer Nature 2008
M. Hoeck, *Cooperation and Technological Endowment in International Joint Ventures: German Firms in China*, Edition KWV, https://doi.org/10.1007/978-3-658-24355-5_1

As the following figure shows, German firms use their technological superiority to successfully serve foreign markets with their products or services.

Figure 1: Germany's World Share for Selected Innovation Indicators (in %)

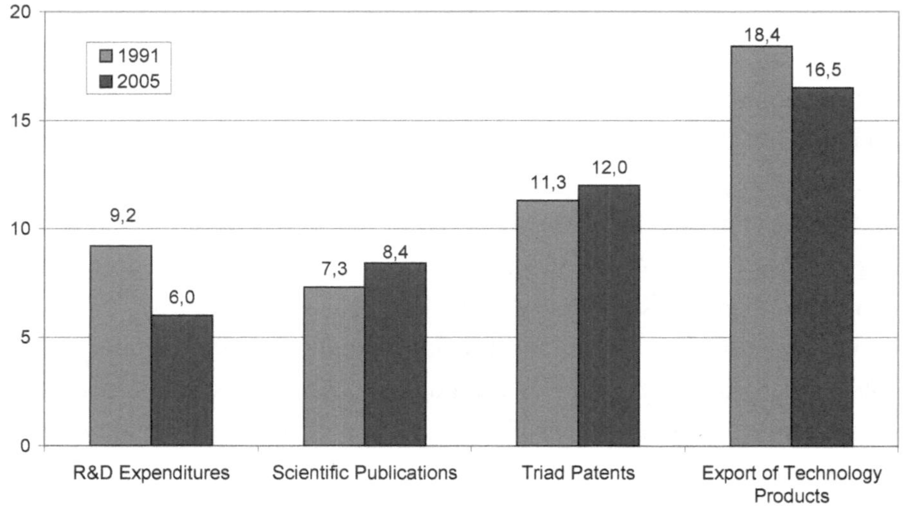

Source: BMBF (2006).

Regarding the target market China, recent growth rates and the large size of the market have triggered a boom of German investment activity. As the following figure shows, German direct investments in 2005 in China accrued more than USD 1.5 bn. This makes Germany the second-biggest western investor after the USA.

Figure 2: Top 10 Sources of FDI to China in 2005 (in USD million)

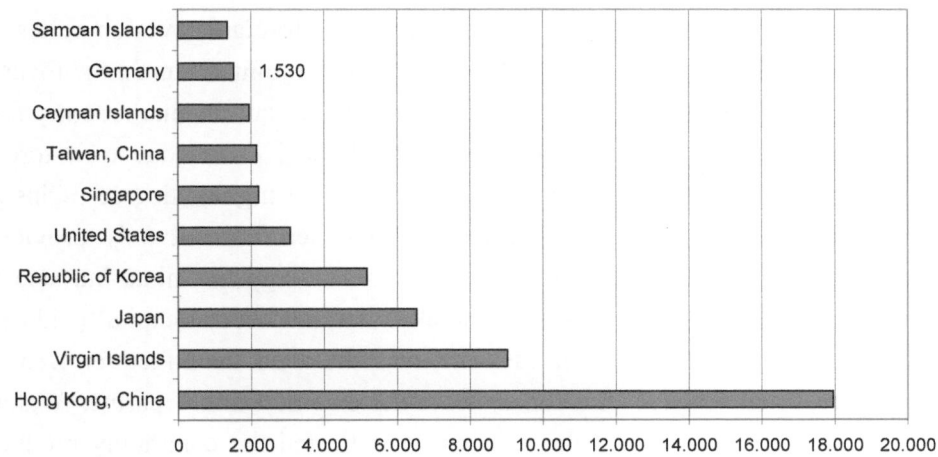

Source: National Office of Statistics of China.

As Holtbrügge and Puck (2005) point out, the total stock of German foreign direct investments in China had already reached about EUR 7.9 bn in 2003, which represents a tenfold increase between 1995 and 2003.

The other side of the coin that has accompanied the recent investment boom in China is the reported or anticipated loss of intellectual property (IP). Due to China's historically weak IP protection regime (World Economic Forum 1995, Huck 2005) and a national culture of imitation (Trempel 2001, Zinzius 2000), market entry to China may result in a foreign investor's diffusion of proprietary technological knowledge. According to a recent study by Roland Berger, two-thirds of international companies engaged in China regard the insufficient protection of patents as a central problem (von Keller et al. 2005).

The Chinese government is working hard to improve the national IP protection system, but is itself known for its policy to absorb technological knowledge by foreign investors (Song 2004, Fu 2005). As a result of the uncertainty regarding IP protection, foreign firms are deterred from deploying technological assets or capabilities to China or even shy away from market entry to China altogether (Geissbauer 1996, Gassmann and Zhen 2004).

The reservations of the foreign investor regarding technology diffusion can result in inter-firm goal incongruence when international market entry is accomplished by international alliance and the foreign investor relies on cooperation with a local partner. In fact, international joint ventures, the focus of this study, maximise the paradox outlined above: IJVs are the market entry form most suitable for transferring complex technological knowledge to a foreign country and are known to give the local partner the highest degree of insight (Root 1994, Inkpen and Beamish 1997). IJVs therefore not only provide a superb channel for German firms to transfer technology to China, but also for Chinese firms to absorb and misuse the absorbed intellectual capital (Tidd and Izumimoto 2002).[2] One recent case involves the food manufacturer Danone, who is reported to have threatened its Chinese joint venture partner Wahaha with a lawsuit because Wahaha apparently replicated the commonly produced food products in separate factories and had sold them in separate distribution channels (Dyer and Maier 2007).

Knowledge diffusion within IJVs can also go the opposite way: local market knowledge gets appropriated by foreign firms and local JV partners are disposed of. Hennart et al. (1995) report the behaviour of Japanese firms in the United States who terminate their JV agreements with local partners, having acquired the necessary local knowledge to run the business on their own. Ono (1991) describes the cases of Ralston Purina, Bayer AG, Monsanto Co., and Sandoz in Japan, who quit international alliances and turned their activities into wholly owned subsidiaries.

As this shows, the issue of knowledge sharing within international joint ventures is not an easy one. Coming back to the point of departure, one thus realises the challenge for German industrial firms when deciding on how to endow international joint ventures in China with technological assets and capabilities. The importance of the technological assets at stake, the high likelihood of knowledge diffusion, and the element of goal incongruence within joint ventures add up to a complex decision framework.

---

[2]   Of course, potential misuse of shared knowledge can go both ways, as will be discussed in detail in the appropriate Chapter of this thesis.

This thesis seeks to shed light on this decision framework by using theoretical insights as well empirical study results from a survey among German-Chinese joint ventures. The next section will specify the research question at hand, taking into account existing literature on the topic.

2. Literature Review and Research Question

Existing literature does provide partial insights for the challenge mentioned above. First of all, there are management insights on how to decide on the value adding activities in target countries such as China (Root 1994, Panhans and Dingeldein 2005, Holtbrügge and Puck 2005, von Keller 2005, Kasperk et al. 2006). This literature tends to reflect on potential IP drain due to foreign activities and proposes appropriate protection strategies. For example, von Keller et al. (2005) propose specific measures to "win the IP game" in China.

Second, there is abundant literature on a foreign investors' choice of the governance structure for value adding activities in the foreign target market – the market entry form (Root 1994). Results suggest that firms who want to transfer sophisticated technology to a foreign market tend to use hierarchical market entry forms such as a joint venture or even a wholly foreign-owned enterprise (Weiss 1996, Fu 2005). This occurs because these market entry forms enable firms to transfer sophisticated or tacit knowledge such as process technologies (Mansfield et al. 1979, Coughlin 1983) or procedures (Katz 1996). Furthermore, very profitable or sensible technologies have been found to be transferred through sole ventures only (Mansfield et al. 1979), because joint ventures carry the risk of knowledge drain to the partner (Niosi et al. 1995). On the other hand, joint ventures are chosen when partners are either complementary in nature (Hladik 1988) or when cultural barriers or government regulation make such an entry form unavoidable (Fu 2005). However, insights such as the ones quoted here do not address the question of how to endow foreign activities with assets, in particular with technological assets and capabilities, *given* that the chosen (or prescribed) governance structure is one that involves cooperating with a local partner.

Third, a growing body of theoretical research (Borys and Jemison 1989, Oliver 1990, Powell 1990, Ring and van de Ven 1992) and empirical research studies (Heide and Miner 1992, Osborn and Baughn, 1990, Seabright et al. 1992, Yan and Gray 1994, Scherer 1995) has addressed inter-firm cooperation within alliances, especially regarding the question of what influences their structure and performance. One work that relates cooperation and the degree of irrevocable investments (such as technological assets) to the structural dimensions of international alliances is Parkhe (1993), using both game-theoretic argumentation and empirical research results.

Fourth, there exist theoretical works that relate to the sophistication of the technology transferred within international technology transfer agreements. Kabiraj and Marjit (1992) outline how the potential threat of entry by the technology receiver shapes the nature of an agreement and determines the 'quality' of the transferred technology. Rockett (1990) investigates international licensing contracts in a model that allows firms to choose the quality of the technology to be licensed along with the structure of payment. Sinha (2001) investigates under which circumstances it is beneficial for multinational enterprises (MNEs) to transfer advanced technology rather than basic technology within international joint ventures. The above quoted works treat technology as a 'black box', however, and refer to technology as an intangible investment that reduces marginal production costs (Kabiraj and Marjit 1992, Sinha 2001).

Finally, this study builds on former studies that try to 'open the black box' of technology by trying to pin down the sophistication of technological assets (The Technology Atlas Team 1987 (a & b), Sharif 1988, Ramanathan 1994, Kahen 1994, Kahen and Griffiths 1995, Cohen 2004) and technological capabilities (Fransman 1984, Dore 1984, Ramanathan 1994, Saha and Nazrul 1998). These works do not relate the degree of technological sophistication to strategic considerations of the technology sender, however.

We therefore partly build on the above quoted strands of literature when investigating the following challenge: German investor firms seeking market entry in China can do so because of technological superiority. However, market entry to China may diminish this superiority if the technological assets that are invested locally can be accessed by a local joint venture partner and possibly diffuse within the industry due to lack of IP protection. Because strategic considerations regarding inter-partner cooperation in international joint ventures lie at the heart of this challenge, we investigate the following two questions:

What factors influence the sophistication of the technological endowment that an international joint venture in China receives from its German parent? and *In what way do strategic considerations regarding inter-firm cooperation and knowledge sharing influence the foreign investor's technology transfer behaviour?*

The next paragraphs will describe how this thesis aims to contribute to the scientific literature regarding these two questions.

3. Structure of the Thesis

The discussion above has shown the theoretical and practical problems relating to the topic of cooperation and technological endowment in international joint ventures. We also presented the specific questions that this thesis is supposed to shed light on.

In order to propose verifiable answers to these questions, we use an empirical research design as described by Martin (1989): after the review of existing theoretical and empirical evidence, we relate to a theoretical paradigm for deriving several testable hypotheses, which in turn are tested empirically. The following figure gives an overview of the logic structure of this thesis.

Figure 3: Structure of the Thesis

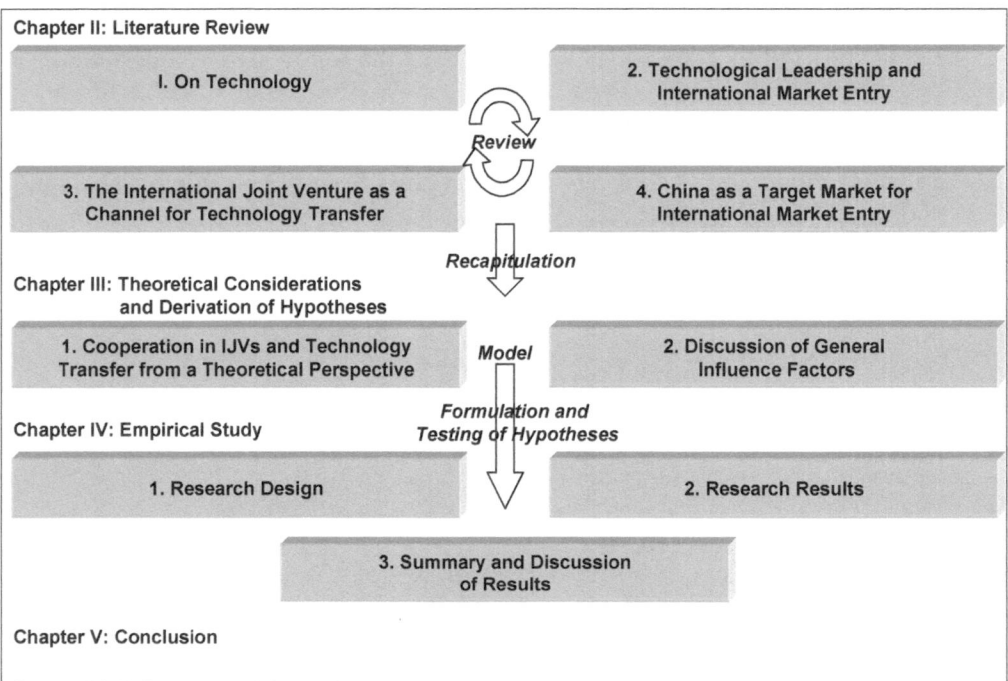

In Chapter II, we present relevant background information on technology and on how technological leadership is related to international market entry. We then review the market entry form international joint venture and how it can be an effective channel for technology transfer. Finally, we discuss the specific circumstances of the target market China, especially regarding foreign investment and technology transfer. This also includes a brief analysis of empirical data regarding technology transfer to China, especially from Germany. We end Chapter II by recapitulating the reviewed insights and describing the research gap that exists.

This thesis contributes to filling this research gap. The starting point is the discussion of theoretical frameworks on cooperation and knowledge exchange and their adaptation to the case of technology transfer to international joint ventures (Chapter III). These theoretical considerations lead to the formulation of several testable hypotheses, which will be presented in Section III-1. Apart from our specific hypotheses, we summarise general factors that influence a foreign investor's endowment of an international joint venture with technological assets and capabilities.

Empirical evidence from a survey among German industrial firms is then used to discuss the validity of the previously derived hypotheses. The design and the results of the empirical study are presented in Chapter IV. Chapter V then offers a conclusion that discusses and summarises the results of this study. Chapter VI contains the reference list as well as several appendices.

Chapter II: Literature Review

This chapter will provide the reader with an overview of relevant literature that relates to the topic of this study. It starts out with a discussion on the term 'technology' and how one can differentiate levels of technological sophistication. Second, the topic 'technological leadership' is reviewed and related to the international market entry of firms. Third, the market entry form 'international joint venture' as a channel for technology transfer is reviewed. Finally, we shed light on the investment conditions in China, focussing on regulations and other conditions regarding technology transfer.

1. On Technology

Technology, the ways to measure it, and the ways to transfer it, are among the key concepts relevant to this study. This is why it is useful to give an overview of the various definitions and concepts at the outset. Additional to the review of relevant concepts given below, Section II-1-2 digs deeper into the question of how to measure the degree of technological sophistication of an organisation, as this degree is an important pillar of the empirical study described in Chapter IV.

1.1. Technology: Definitions and Characteristics

In this chapter, we will review the various concepts and definitions related to the term 'technology'. After presenting some definitions for technology, we give an overview of the characteristics of technology that are relevant for this study.

1.1.1. Definitions

On a very broad level, technology can be described as the principal means adopted by nations seeking developmental progress and higher standards of human life (Cohen 2004). In economic growth theory, technological progress is thus treated as the residual factor explaining the growth of a nation's GNP when

© Springer Fachmedien Wiesbaden GmbH, part of Springer Nature 2008
M. Hoeck, *Cooperation and Technological Endowment in International Joint Ventures: German Firms in China*, Edition KWV, https://doi.org/10.1007/978-3-658-24355-5_2

one controls for all other input factors (such as capital and labour) over time (Solow 1956, Mankiw 1997).

Of course, opening up the 'technology black box' requires a more detailed view of what comprises technology. For this reason, we cannot treat technology as the residual, but have to look at its characteristics, and here the variety of definitions begins. A simple starting point for defining technology is to describe it as the practical knowledge of how to do and make things (Litter 1988). Narrow definitions of technology refer mainly to the technical aspects of technology and regard technology as a means for transforming natural resources into produced resources (Cohen 2004). Technology is then understood as the actor transforming certain inputs into desired outputs (Goh 2004). According to the Technology Atlas Team (1987a), any such transformation may be described in terms of inputs, outputs, production activity and technology. The respective framework is depicted below.

Figure 4: Technology and the Production Process

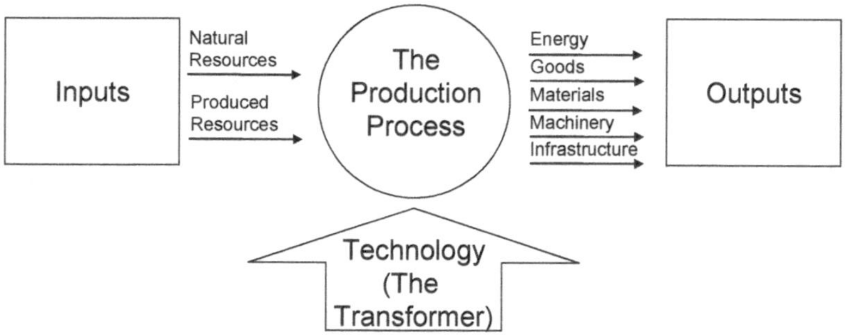

Source: Cohen (2004), p.90.

Technology cannot be reduced to machines (Ernst and O'Connor 1989). It has to do with certain kinds of knowledge, which allow the adaptation of means to ends. Part of this knowledge is embodied in machines, but most of it is not. It is embodied elsewhere: in the brains of people, in organisational structures and in behavioural patterns. Technology is thus usually described as involving many elements, such as engineering, organisational know-how, and economic, societal, and managerial factors (Goh 2004).

Maskus (2003, p.14) proposes to define technology as "the information necessary to achieve a certain production outcome from a particular means of combining or processing selected inputs." Another proposition by Cohen (2004, p.66) is to define it as a "set of specialised knowledge applied to achieving a practical purpose". There is clearly no scarcity of definitions. One common characteristic among most definitions is that technology contributes to the productivity with which outputs are generated from inputs and to the market value of those outputs (Maskus 2003). Cohen (2004) thus regards technology as a *capability* that can aid in accomplishing a certain task, and that comes in the form of physical structure or knowledge embodied in artefacts (software, hardware, or methodology).

Building on the definitions presented above, we now introduce some relevant characteristics of technology, before we elaborate on the question of what constitutes 'sophisticated technology' (Section II-1-2).

1.1.2. Characteristics of Technology

Technology has some very specific characteristics that should be taken into account for any analysis. These are described in the following paragraphs.

One important characteristic that has already been hinted at is that technology may come in *various forms*. A variety of classifications for these forms have been devised over time. Scholars broadly distinguish between technology in the form of plants, machinery, equipment, or documents – 'hard' technology – and technology that is embodied in organisations via the skill of its people, its organisational routines, or even its cultural adaptation to a certain environment – 'soft' technology (Goh 2004). Stewart (2002) classifies the components of technology as (1) products, (2) machines and processes to make the products and (3) the software necessary for the production of the products. Commonly, one also uses the distinction between 'codifiable' and 'tacit' knowledge. Codifiable or 'explicit' knowledge is knowledge that can be codified into text or diagrams. Tacit knowledge on the other hand is subconsciously learned from experience and can be very difficult to articulate and codify. One example for tacit knowledge is a complex production process embedded in an organisation (Nonaka 1991, Kogut and Zander 1993, Zack 1999, Maskus 2003, Dong 2004).

Another property of technology is its *public good character*. Technology bears the two characteristics of a (pure) public good – non-rivalry and non-excludability (Maskus 2000). Non-rivalry means that the consumption of a good by one individual does not reduce the amount of the good available for consumption by others. Non-excludability means that it is not possible to exclude individuals from a good's consumption. Because technology development is generally costly, innovative firms have an interest in maintaining excludability in order to generate a market return on R&D investments. Excludability is thus the essence of intellectual property rights, which attempt in principle to balance the returns of innovation against the needs for diffusion (Maskus 2003).

A final property that needs to be pointed out is the fact that trading with technology is complicated by *asymmetric information*. The essential problem is that the owner of a technology may have complete knowledge about its specifications, its effectiveness when deployed under different circumstances and associated know-how, while the buyer has far less information about it (Maskus 2003). This impedes technology trade because the seller is unwilling to reveal the information before the buyer has confirmed the purchase. On the other hand, the buyer is not willing to buy a certain technology without having the possibility of prior testing (Maskus 2000).

The next paragraph will introduce relevant literature on the relationship between technology types and industry classifications.

1.1.3. Technology Types and Industry Classifications

Technology is concerned with the production of industrial outputs, which is why technology classifications often relate to end products, industries, or the production process. As has been pointed out by Hamel (1991), companies that belong to a certain industry usually share a certain proportion of their technological basis, as the products they produce require similar production processes. On the other hand, a single technology may be used to serve a variety of markets. One may thus argue that technology is not only inextricably linked to the production of products or services, but it may even be the defining property for firms when asking for its 'core competencies' (Hamel and Prahalad 1990, Dosi et al. 1992).[3]

A classification of industries according to the technologies they employ has been offered by Engelsmann & van Raan (1994). They order technologies according to the field in which corresponding patents are filed. The result shows a certain relationship between technologies and industries (see Appendix 1).

Woodward (1958, 1965) already relates to the industry-defining characteristics of a (production) technology in her classification system from 1965. As shown in Table 1, the main classes distinguished by her are 'unit and small batch production', 'large batch and mass production', 'process production', and 'combined systems'.

---

[3]   A firm's core competence has been defined as the collective learning in the organisation, which creates the ability to consolidate corporate-wide technologies and production skills into competencies that empower individual businesses to adapt quickly to changing opportunities. Moreover, a core competency is something that a firm can do well and that meets the following three conditions: (1) it provides customer benefits (2) it is hard for competitors to imitate and (3) it can be leveraged widely to many products and markets (Prahalad and Hamel 1990, p.82). Dosi et al. (1992) define a core competence as a set of differentiated technological skills, complementary assets, and organizational routines and capacities.

Table 1: System for Classifying Technologies

Unit and Small Batch Production
1. Production of units to customers' requirements
2. Production of prototypes
3. Fabrication of large equipment in stages
4. Production of small batches to customers' orders
Large Batch and Mass Production
5. Production of large batches
6. Production of large batches on assembly lines
7. Mass production
Process Production
8. Intermittent production of chemicals in multipurpose plants
9. Continuous flow of production of liquids, gases, and crystalline substances
Combined Systems
10. Production of standardised components in large batches, subsequently assembled diversely
11. Process production of crystalline substances, subsequently prepared for sale by standardised production methods

Source: adapted from Woodward (1965), p.39.

The important idea is that companies employing similar production technology face rather similar challenges or tend to be organised in a similar fashion, setting them apart from companies that employ other technologies. Based on her categorisation Woodward (1965) not only concludes that distinct relationships exist between the technology classifications and the subsequent structure of a firm, but also that the effectiveness of the organisation is related to the 'fit' between technology and organisation structure.

A more general approach has been taken by Thompson (1967). Referring to the work of Woodward, he defines a broader classification of technologies based on the principal logic of interdependence of activities within an organisation. These resulting three types of technologies are 'long-linked' technology (sequential interdependence of tasks), 'mediating' technology (pooled interdependence of tasks) and 'intensive' technology (reciprocal interdependence of tasks). Thompson's argument is that each type of technology imposes specific requirements and thus influences the structure of the whole organisation.

The considerations that relate technology to production processes and eventually the structure of organisations are important considerations in the context of this study. Specifically, we will further investigate the relationship between the technology to be employed in a foreign market and the chosen market entry form (Section II-2-2). First, the next section discusses possible indicators for the degree of sophistication of an organisation's technological endowment.

1.2. Indicators for Technological Sophistication

When trying to describe the sophistication of a technology employed by an organisation, research has generally drawn on one of the four following indicators:

- the functional activities covered by an organisation,
- the resources with which an organisation is endowed,
- the capabilities of that organisation or
- the performance of that organisation.

We will briefly elaborate on each of these approaches below, as they form an integral component of the empirical study described in Chapter IV.

1.2.1. Which Functional Activities are Carried Out?

In a simple way, one can assess an organisation's technological sophistication by recording the *functional activities* that are covered by it. Recording the functional activities covered by an organisation – for example using the value chain model of Porter (1985) as a reference – can result in information regarding its technological sophistication. For reference, the value chain model is reproduced below.

Figure 5: Value Chain Model

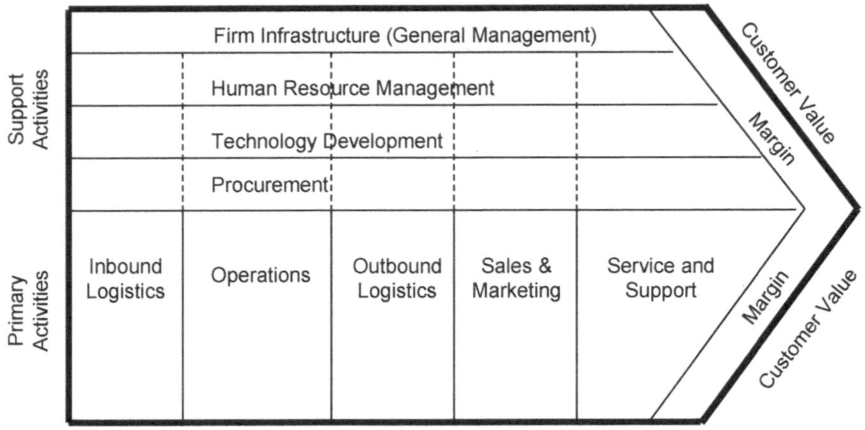

Source: Porter (1985), p.37.

Functional capabilities of an international firm's foreign operations can vary a lot. As will be described in later sections, some foreign operations of international firms only have representative function or are used for raw material sourcing. These are tasks that do not require technological skills regarding the company's end products. Other organisational entities carry out application development or even research tasks and can thus be assumed to have very strong technological capabilities (Beckmann and Fischer 1994). Carrying out the service function, the human resource function or the procurement function also requires knowledge about the product technology to be employed to a certain degree. As Beckmann and Fischer (1994) emphasise, the function 'technical sales' comes close to the function of 'application development', as products and services have to be adapted according to customer requirements. It thus presupposes technological understanding of the products or services sold.

One can conclude that a general impression of a company's technical capabilities can be inferred from looking at the functional activities covered. Many empirical studies (e.g. Fu 2005, Weiss 1996) record the functional tasks of organisations with 0/1 variables and argue that if a certain organisation conducts research, it will have higher technical capabilities than one that does not. However, this approach is rather simplistic, and there exist more differentiated approaches to assess an organisation's technological sophistication. These are the subject of the following paragraphs.

1.2.2. The Sophistication of an Organisation's Technological Resource Endowment

A variety of scholars have described ways to assess an organisation's technological sophistication by identifying its endowment with technological resources. One widely-accepted approach (The Technology Atlas Team 1987a&b, Sharif 1988, 1995, 1997, Ramanathan 1994, Kahen 1994, Kahen and Griffiths 1995, Cohen 2004) is to divide technology into four components, namely 'technoware', 'inforware', 'humanware', and 'orgaware'. These components are defined as follows:

1. Technoware, or *object-embodied technology*, refers to the tangible and palpable part of the machineries that an organisation is endowed with. It comprises physical equipment such as plants, machines, or even single tools.
2. Inforware, or *document-embodied technology*, refers to the accumulated knowledge about a technology that can be documented, such as documentation regarding the production process and related background information.
3. Humanware, or *person-embodied technology*, refers to the human resources that an organisation can draw on to realise the potential of technoware, such as the skills and experiences of people engaged in the organisation.
4. Orgaware, or *institution-embodied technology*, refers to the supporting net of principles, practices, and arrangements that govern the effective use of the other three components. It is also related to the size of an organisation.

These four components form an integrated set, and no technical or transformational operation can take place in the absence of any of these components (Sharif 1997, Cohen 2004). For example, even the most modern machinery is of no use if an organisation has a lack of qualified personnel to operate that machinery. Thus, sophisticated technology usually draws on all four of the above-mentioned components simultaneously.

This framework has been used for the assessment of technological sophistication or technological advancement. The argument is that the overall level of an organisation's technological sophistication can be described by assessing the level of sophistication of each of the four components – technoware, inforware, humanware, and orgaware – and the improvement of any of the four components implicitly leads to the improvement of the whole set (The Technology Atlas Team 1987a&b, Sharif 1988, Cohen 2004).

The next table shows different levels of technological sophistication for each of the interacting components with corresponding examples.

Table 2: Levels of Sophistication of the Components of Technology

	Level of Sophistication	Examples
Technoware		
1.	Manual tools	Screwdriver, hand drill
2.	Powered equipment	Grinder, power drill
3.	General purpose facilities	Milling machine, lathe
4.	Special purpose facilities	Textile power looms, airjet weaving loom
5.	Automatic machines	Soft drink bottling plant
6.	Computerised facilities	Numerical Control (NC) machines
7.	Integrated facilities	Completely robotised assembly plants, integrated plants
Inforware		
1.	Familiarising facts	Brochure, images
2.	Describing facts	Technical booklet, process description
3.	Specifying facts	Performance and usage specifications
4.	Utilising facts	Standard operating and maintenance manuals
5.	Comprehending facts	Process theories, design data and calculations
6.	Generalising facts	Development information generated through indigenous R&D
7.	Assessing facts	Comprehensive information on the latest developments
Humanware		
1.	Operating abilities	Unskilled and semi-skilled operators
2.	Setting-up abilities	General technicians, skilled operators
3.	Repairing abilities	Special technicians, maintenance engineers
4.	Reproducing abilities	Production engineers
5.	Adapting abilities	Design engineers
6.	Improving abilities	Development engineers (development)
7.	Innovating abilities	Development engineers (research)
Orgaware		
1.	Individual linkages	Small firm
2.	Collective linkages	Connected small firms
3.	Departmental linkages	Small-scale organisation
4.	Enterprise linkages	Medium-scale organisation
5.	Industrial linkages	Large-scale organisation
6.	National linkages	Multi-location organisation
7.	Global linkages	Transnational organisation

Source: Adapted from Cohen (2004), p. 94.

The four dimensions technoware, inforware, humanware and orgaware can be used for a comprehensive assessment of the sophistication of an organisation's technological resource endowment. For this reason, the empirical study described in Chapters IV and V relies on it. A discussion of how the approach has been implemented can be found in Section IV-1-2.

The next section will introduce another view of technological sophistication, which is to assess the general skills related to an organisation's technological *capabilities*.

1.2.3. The Sophistication of an Organisation's Technological Capabilities

A view on technological sophistication of an organisation that can be regarded as complementary to the resource view described above is the identification and assessment of an organisation's technological *capabilities*. Many attempts have been made by different authors to define the term 'technological capability' and measure it – typically in the context of technology transfer to development countries (Saha and Nazrul 1998, Kahen 1994, Kahen and Griffiths 1995).

One assessment scheme for technological capabilities offered by Fransman (1984), one of the pioneers in this area, distinguishes between the abilities to:
- *search* for available alternative technologies and to select the most appropriate technology for importation,
- *master* imported technology and successfully use it for the transformation of inputs into outputs,
- *adapt* imported technology in order to suit local production conditions,
- further *develop* the adapted technology as a result of local incremental innovation,
- *institutionalise* the search for more important innovations and breakthroughs with the development of local R&D facilities and
- *upgrade* a technology by carrying out basic research.

Dore (1984), on the other hand, understands technological capability as the combination of three kinds of independent capabilities, namely an independent

21

technology *trend monitoring* capability, an independent technology *learning* capability; and an independent technology *creation* capability.

A third approach has been suggested by the World Bank (1985), subdividing technological capability into the following three capabilities:
- *Production* capability, consisting of production management, production engineering, maintenance of capital equipment, and the marketing of produced output;
- *Investment* capability, consisting of project management, project engineering, procurement capabilities, and manpower training; and
- *Innovation* capability, describing the ability to create new technologies and bring them to economic use.

As the frameworks above suggest, technological capabilities involve both elements of the 'absorption, digestion, and improvement' of technology, as well as the handling of technology in different functional areas of an organisation. A useful framework should take account of these different elements and clarify its interactions.

Such a framework has been provided by Ramanathan (1994), who suggests the following four categories:
- *Operative capabilities*: Ability to operate and control plant and equipment, plan and control production activities, provide information support and networking for operations, and maintain the plant and equipment in good order.
- *Acquisitive capabilities*: Ability to carry out a detailed engineering study, independently search for good technology sources, assess technologies, decide on the technology transfer mode, and negotiate the terms of a technology acquisition.
- *Innovative capabilities*: Ability to duplicate acquired technology, adopt and carry out improvements in imported technology, and carry out an independent technology development plan.
- *Supportive capabilities*: Ability to undertake project planning and execution, obtain funds for prototype development and modernisation, plan and implement human resource development, and identify and develop new markets for the firm's existing and new products.

While all of the above-described typologies are equally useful, the latter one particularly is not only comprehensive, but also takes account of the interactive nature of technological capabilities.[4] It is therefore the typology of choice for this study and will be referred to again as part of the research design description (Section IV-1).

1.2.4. The Performance of an Organisation as a Proxy for its Technological Skills

A final approach on how technological sophistication can be assessed is to use an organisation's performance as a proxy for its technological skills. As Argote and Ingram (2000) describe, technological learning manifests itself through changes in the performance of an organisational unit.

In economic models, it is common to describe technology as a main factor influencing the marginal costs of production. Improvements of technology due to innovation or technology transfer are then represented by a decrease in marginal costs (Kabiraj and Marjit 1992, Sinha 2001). A particular challenge in assessing technology through measuring performance is to control for factors in addition to technology may affect the performance of an organisation (Argote 1999). The approach to derive technological sophistication by looking at performance data is therefore limited to specific applications where factors influencing a unit's performance can either be controlled for or are very limited.

For this reason, we do not use the performance-based approach in our own study and thus the discussion is kept relatively brief. We restrict ourselves to mentioning a few studies using this approach.
A first example is a study conducted by Epple et al. (1991). They estimate unit costs in production plants (specifically direct labour hours per unit) and track what happens when a second shift gets introduced. They find that although the second shift does not operate as efficiently as the first one immediately, it starts

---

[4]   As Saha and Nazrul (1998) point out, the approach by Ramanathan (1994) understands the individual components of technological capability to be multiplicative while other frameworks, such as the one presented by Fransmann (1984), describe how capabilities build on each other "additively".

out more efficiently and experiences faster learning than the initial shift –
concluding that significant knowledge transfer takes place.

A second example is the study of Darr et al. (1995). They estimate the extent to
which the productivity of fast-food stores is affected by the experience of the
other stores in their franchise.

A final example is Benkard (1999), who analyses the extent to which the
experience of producing one aircraft model affects the amount of labour
required to produce a subsequent model, using production data from the aircraft
manufacturer Lockheed.

The cases above are three examples for cases where a change in performance
can be traced back to technological learning. Marginal cost levels can then be
used to express the technological sophistication of an organisation.

The different approaches to assessing and comparing technological
sophistication are a crucial pillar of the empirical study described in Chapter IV.
Especially the discussion in Section IV-1-2 that describes possibilities for
measuring the dependent variable picks up on the arguments mentioned here.

In the following paragraphs, however, we review literature on technology
transfer between organisations, in particular between organisations that stem
from different countries.

1.3. International Technology Transfer (ITT)

This section highlights some concepts related to (international) technology
transfer (ITT) that are relevant in the context of this study. Because the term
'technology transfer' has slightly different connotations in different contexts
(Cohen 2004), it is necessary to define it in the context of this study: the transfer
of (advanced) technology from a German industrial firm to an international joint
venture in China.

Second, it is useful to understand how technology can be transferred in general.
Not only do different mechanisms of technology transfer exist, but each of these
mechanisms entails implications for the successful management of technology
transfer. It is relevant to look at these implications, because the technological

endowment of a given organisation after technology transfer has occurred is not only a result of the *willingness* of the parties involved, but also to a significant extent a result of *feasibility*.

1.3.1. ITT in the Context of this Study

It is often argued that the term 'technology transfer' is complicated and confusing, leading to frequent misunderstandings regarding the nature of the process of transfer of technology (Odedra 1990). Thus, several definitions for technology transfer may be found in the literature.

The definition problem arises because the term involves two complex and multidimensional concepts: 'technology' and 'transfer' (Cohen 2004). Indeed, the mere interpretation of the term 'transfer' makes a large difference. According to the *Oxford English Dictionary* (4th edition), the term 'transfer' could mean 'to move from one place to another' or 'to hand over the possession of property or right'. Only the latter definition presumes the purposeful action of a technology 'sender' and is better described by the term 'transmission' (Dunning 1981).

According to Maskus (2003), technology transfer generally refers to any process by which one party gains access to a second party's information and successfully learns and absorbs it into his production function. This definition is rather broad, because it leaves room for technology transfer that occurs between willing partners in voluntary transactions, knowledge flows between different units within a single organisation, and even 'non-conventional' channels such as technology diffusion by reverse engineering.

Definitions that presuppose an active transmission of knowledge have been offered by Cohen (2004) and Dong (2004). Cohen (2004) defines technology transfer as the systematically organised exchange of information between two enterprises, generally between different countries. Dong (2004) describes technology transfer as a process in which technology and related knowledge is made available by a sender to a receiver. Indeed, he emphasises that technology transfer mainly is concerned with the organisational process involved, and not so much with the 'movement' of technology.

25

Building on the two definitions above, we adopt the following working definition: technology transfer is a goal-oriented, planned transmission of technology from a sender to a receiver and can include the transfer of material as well as immaterial technology.

We now further specify the type of technology transfer that this study is concerned with. For this purpose, we discuss the following dimensions that can be used to distinguish between different types of technology transfer (Paquin 2000, Maskus 2003):

- Technology transfer through market vs. non-market channels
- Direct vs. indirect technology transfer
- Technology transfer with capital investment vs. without capital investment
- Intra-organisational vs. inter-organisational technology transfer
- Public vs. private technology transfer

We will briefly discuss each of these distinctions and then pin down the exact nature of the technology transfer to be described in our study.

A first relevant distinction is between *market* and *non-market* channels of technology transfer. Market channels are characterised by a technology sender's intention to pass on technological capabilities. Formal transactions underlie the technology movement. Non-market transfers such as imitation are either not actively encouraged by the sender or even occur without the sender's consent (Root 1994, Maskus 2003). The following table enumerates the most often quoted market and non-market channels between firms. As can be derived from the table, international joint ventures are generally regarded as a market channel, although technology flows within IJVs may be non-anticipated and/or imitative (see Section II-4).

Table 3: Market and Non-Market Channels of ITT between Firms

Market Channels:
Published material (journals, books)
Trade in goods and services (especially machinery or intermediate goods)
Licensing or collaborative agreements
Foreign direct investment (incl. joint ventures)
Cross-border movement of personnel
Non-Market Channels
Imitation / Reverse engineering
Departure of employees
Data in patent applications and test data
Temporary migration

Source: Adapted from Maskus (2003, p.15) and Cohen (2004, p.113)

Second, a related distinction is the question of whether technology transfer is *direct* or *indirect*. Direct technology transfer is the result of the sale of a certain technology via the market. On the other hand, indirect technology transfer occurs as a means to reaching a certain goal when capital investments are undertaken. This tends to be the case in international joint ventures, because technology senders who are foreign market investors usually pursue goals such as market entry or the optimisation of cost structures, rather than the sale of technology for its own sake (Bennett et al. 2001).

Third, it is of relevance whether technology transfer is or is not accompanied by *capital investment* (Paquin 2000). Foreign direct investments (including equity joint ventures) are examples for capital investments leading to technology transfers, whereas licensing contracts or contractual joint ventures do not entail capital investments in the target country.

A further relevant distinction is whether technology does or does not cross organisational boundaries (Paquin 2000). Here exists a continuum of possibilities. At one extreme, technology transfer can be the result of an arm's length transaction between two unrelated firms (such as the sale of a licence). At the other extreme, technology may be transferred within organisational boundaries (e.g. to a foreign, wholly owned subsidary). The described case of transferring technology to an international joint venture is something in between, and the implications of this fact are manifold (see Section II-4).

Finally, technology transfer projects engaged in by private parties can differ significantly from those undertaken by public authorities. The former activities represent commercial, profit-seeking projects between firms, whereas the latter usually represent initiatives for the sake of development support (Maskus 2003). In this work, we are only concerned with private technology transfer transactions, at least with respect to the German industrial firms that 'send' the technology.

Summarising the considerations above, we are now in a position to narrow down the types of technology transfer to be considered in this study. Technology transfers from German industrial firms to international joint ventures in China take place within the private domain and are accompanied by capital investment. International joint ventures represent a market channel for technology transfer, although not all of the technology might be directly intended by the sender. Only the question of whether the transfer is intra- or inter-organisational is difficult to answer.

A more detailed discussion of technology transfer within international joint ventures is offered in Section II-3. Subsequently, we briefly introduce some insights and challenges regarding the process of technology transfer.

1.3.2. Relevant Issues in ITT Management

A basic model that can be used to highlight some common insights and challenges for the management of technology transfer is shown below (Chen 1996).

Figure 6: A Basic Model of the Technology Transfer Process

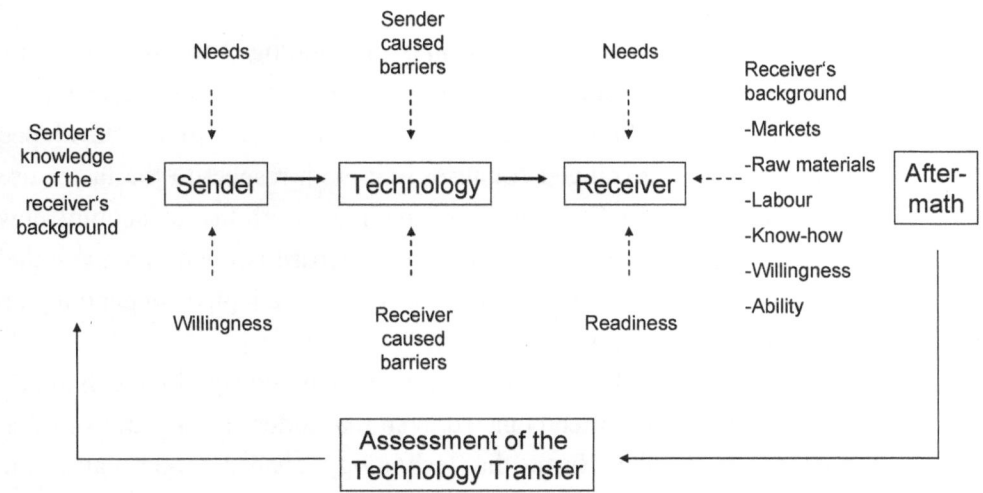

Source: Chen (1996), p.13.

Irrespective of how technology is transferred and for what purpose, there are always two parties involved (the sender and the receiver), and each of these parties has its specific background. The successful accomplishment of technology transfer not only relies on each party's willingness and capability to engage in the transaction, but also in the common ability to surmount certain barriers inherent to ITT projects.

In this context, we emphasise some barriers to technology transfer that have been frequently cited in previous literature. These are, among others:

- *Costs*: Transfer costs can reach significant – if not prohibitively high – levels in international technology transfer projects. In a survey of 26 ITT projects, technology transfer costs have made up between 2% to 59% of total project costs (Teece 1977).
- *Communication problems:* Because human interaction is the key factor determining the success or failure of an ITT project (Al-Obaidi 1999), communication problems can be significant barriers. As argued by Rogers (2003), effective communication regarding sophisticated technology usually only works between 'similar' people.[5] For the case of China, differences in language and culture have been frequently quoted as

---

[5] Rogers (2003) here uses the term 'homophily'.

significant barriers to effective knowledge transfer (Gassmann and Zhen 2004).

- *Matching supply with requirements:* Technology transfer is only successful if the appropriate technology is being transferred. Requirements for technology generally differ between firms in developed countries and less-developed countries, e.g. with respect to the factor use (Stewart 2002). On a firm level, the recipient's ability to acquire new knowledge depends on current levels of expertise (Cohen and Levinthal 1990). Therefore, technologies may have to be adapted, depending on who is going to use them.

- *Goal incongruence:* The transaction partners who engage in a technology transfer agreement are often engaged in similar industries. By transmitting technology to another firm, the technology sender might cause the emergence or strengthening of competition (Lin and Saggi 1999, Kabiraj and Marjit 1992, Goh 2004, Feess et al. 2006), a situation that Rouach (2003) calls the 'boomerang effect'. Technology donors thus try to restrict the extent of technological capability of the technology receiver. This issue is particularly emphasised in cases where technology recipients are located in low-cost countries ('North-South' technology transfer debate, see e.g. Frame 1983, de Almeida 1995, Ming 1996).

Relating to the challenge of goal incongruence mentioned above, we point out that technology transfer projects evolve through typical phases. Rouach (2003) describes these phases as follows:

- *Incubation:* Both parties evaluate their technological capabilities, human abilities, and know-how. This phase also includes the evaluation of the competitive environment.

- *Anticipation:* Both parties anticipate the mode of transfer, the choice of potential partners and clients, and agreements like IP protection or the pricing of technology.

- *Confrontation:* The parties negotiate the terms of the agreement. This phase is a period of aggressive bargaining. At this stage, existing cross-cultural concerns and goal incongruence become evident.

- *Implementation:* After both parties have reached an agreement, they are concerned with financing, project management, adaptation of technologies to specific local requirements and the actual transmission of know-how.

- *Follow-up:* Finally, partners are involved in monitoring the developments, collecting / paying royalties, and potentially preparing follow-up steps with regard to the project in question.

The above description shows that the motivations and incentives for each party subject to a technology transfer project change during the course of a project. This adds an additional challenge for the successful completion of a technology transfer project.

In summary, we point out that the process of technology transfer is complex and the end result (the level of technology employable by the receiver) depends on a great variety of factors. These are important considerations to be taken into account for the quantitative study described in Chapter IV.

Next, we summarise insights on how technological leadership relates to international market entry.

2. Technological Leadership and International Market Entry

The previous chapters have provided a brief overview of technology in general, technological sophistication and topics related to technology transfer. These are important basic considerations that serve as building blocks for the quantitative study described in Chapter IV.

The topics that we are going to investigate in this section are a) explanations for the international market entry of firms; b) why technological leadership is of relevance for the international competitive advantage of firms; and finally c) the relationship that exists between a chosen market entry form and the extent of technology transfer implicit in a market entry strategy.

It will be shown that the transfer of technology is often a necessary condition to be successful in foreign markets and that some forms of market entry (such as the international joint venture) are especially adequate for such a transfer.

2.1. The Role of Technological Leadership for International Market Entry

International market entry – the expansion of a firm's activities into foreign countries – can most simply be viewed as growth across borders, and a necessary condition for international expansion is a firm's competitiveness in foreign markets (Root 1994). In this section, we review common explanations for the international expansion of firms. We then describe how technological leadership and the transfer of technological assets relate to a firm's competitiveness in international markets.

2.1.1. Explanations for International Market Entry

Researchers have proposed many different explanations of international market entry. A good review of the various theories and their proponents is provided by Weiss (1996). The following table shows an overview in order to provide a basis for further discussion.[6]

---

[6]   An extensive discussion of the literature shown in the table would go beyond the scope of this thesis, and the reader shall be referred to the review provided by Weiss (1996).

Table 4: Theories for the Explanation of International Market Entry

Theories of Industrial Organisation		Theories of the Firm/MNC	
Theory	Author	Theory	Author
Capital Theory	Nurkse (1934)	(International)	Coase (1937)
	Heidhues (1969)	Transaction Costs	Williamson (1975, 1985)
			Hennart (1982, 1988)
Monopolistic	Bain (1956)		Teece (1981, 1986)
Theory	Hymer (1960)		
	Kindleberger (1969)	Property-Rights	Alchian (1965)
	Johnson (1970)		Alchian/Demsetz (1973)
	Caves (1971,1974)		
		Internalisation	McManus (1972)
IO & Strategy	Porter (1980, 1985)		Buckley/Casson (1976)
	Oster (1990)		
		Behavioural Theory	Aharoni (1966)
Product Life Cycle	Vernon (1966, 1974)		
	Wells (1972)	Diversification	Rugman (1975)
			Scherer (1975), Wolf (1977)
Parallel Conduct	Knickerbocker (1973)		
		Eclectic Theory	Dunning (1977, 1979, 1980)
Learning	Johanson/		
	Wiedersheim-Paul (1975)	Evolutionary Theory/	Collis (1991)
	Johanson/Vahlne (1977)	Resource-Based View	Kogut/Zander (1993)

Theories of International Marketing		Theories of International Trade	
Theory	Author	Theory	Author
Application of	Anderson/	Absolute/Comparative	Smith (1776)
Economic Theories	Gatignon (1986)	Cost Advantages	Ricardo (1817)
	Hill/Kim (1988)		
		Factor	Heckscher (1919)
Globalisation	Buzzell (1968), Levitt (1983)	Proportions	Ohlin (1933)
	Porter (1986), Ohmae (1985)		
		Neo Factor	Kenen (1965), Kravis (1956)
Market/ Product	Ayal/Zif (1979)	Proportions	Leontief (1969)
Diversification			
		Monopolistic	Posner (1961)
Business Strategy	Root (1994)	Advantages	Linder (1961)
Approach	Prahalad (1975)		Grubel (1967)

Source: Adapted from Weiss (1996), p. 16.

Summarising the large amount of work quoted above, we refer to Collis (1991), who argues that literature relating to international market entry has produced a consensus view that is best captured in four postulates.

The first postulate is that international market entry is only relevant whenever there are important interdependencies between competitive positions in different countries. The acid test is whether a firm is 'better off' in country A by virtue of its position in country B (Collis 1991).

Second, there is consensus of the sources for competitive advantage that an internationally operating firm can draw on. For example, companies pursuing a low-cost strategy benefit from scale economies due to size (Porter 1985) or organisational efficiencies inherent to being a large producer on a global scale

(Bartlett and Ghoshal 1998). On the other hand, companies pursuing differentiation strategies may benefit from a global brand name or scale and scope economies in its R&D programme (Porter 1985, Collis 1991).

Third, the firm's ability to realise benefits from global operation while minimising administrative costs will depend on the optimal location of each of the firm's activities, and how these activities are controlled (Dunning 1997).

Finally, the insight that an organisation's structure should be aligned with and derived from the strategy (Chandler 1962) also holds for the international case. In particular, the balance between localisation (differentiation) and integration (global efficiency) must be consistent with the chosen strategy (Prahalad 1975, Bartlett and Ghoshal 1998).

These four insights are a very simplified summary of insights from the literature relating to international market entry, but they do indicate how technological leadership is being taken account of. One the one hand technology is been regarded as a source of monopolistic advantage and thus the ability for foreign market entrants to compete with domestic incumbents. Examples for theories that take account of this fact are the monopolistic theory, the product life cycle theory, the resource-based view and the eclectic theory.

On the other hand, technology has been identified as one capability that needs to be transferred to a foreign national market for successful competition. It thus affects the way how an international firm competes in a foreign country. Examples for theories that take account of this aspect of technology are the international application of the transaction cost theory, the internalisation theory, and the eclectic theory.

It can thus be argued that the issue of specific internal assets, especially technology, has always received much attention by scholars of international firm activity. However, only one strand of literature, namely the proponents of the resource-based view and its application to international market entry, has emphasised technology as the main issue in international market entry. The following paragraphs therefore give an overview of the resource-based view in general. We then focus on the application of the resource-based view to international market entry.[7]

2.1.2. The Resource-Based View

The dominating view held in the 1980s was that the profitability of a company is mainly determined by industry structure ('market based view', see e.g. Caves and Porter 1977, Porter 1980, 1985). For example, the 'five forces model' by Porter (1985) describes the attributes of an attractive industry and thus suggests that all firms in that industry will exhibit similar levels of firm performance. Implicitly, this approach adopts the simplifying assumption that firms within an industry are identical in terms of the strategically relevant resources they control and the strategies they pursue (Porter 1981, Rumelt 1984, Scherer 1980).

A counter-position has been adopted by proponents of the resource-based view. The theoretical approach here is not to see the firm through its activities in the product market, but as a unique bundle of tangible and intangible resources (Wernerfeldt 1984). A firm's above-average profitability is argued to stem from specific and idiosyncratic attributes (Barney 1991, Bamberger and Wrona 1996, Macharzina 2003).[8] It is then internal resources instead of the industry structure that are responsible for maintaining a company's sustainable competitive advantage in (international) markets.

As the figure below suggests, assets have to meet certain criteria in order to contribute to a firm's sustained competitive advantage: they have to be valuable, rare, imperfectly imitable, and non-substitutable (Dierickx and Cool 1989, Barney 1991, Welge and Holtbrügge 2003).

---

[7]   Kogut and Zander (1993) here propose the name 'evolutionary theory of the multinational corporation'.

[8]   Firms are assumed to be idiosyncratic because they accumulate different physical and intangible assets throughout their history (Dosi et al. 1992, Barney 1991).

Figure 7: The Resource-Based View of the Firm

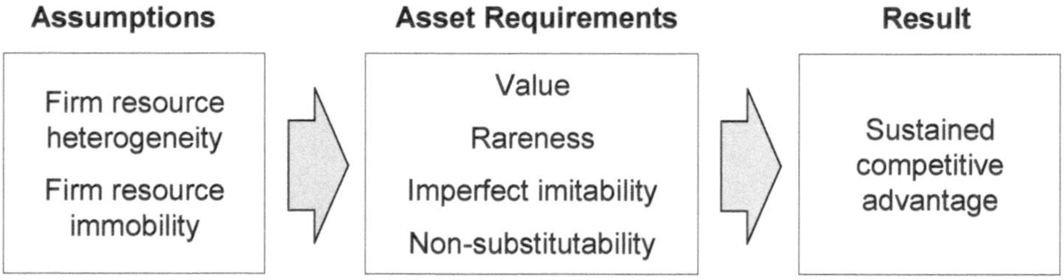

Assumptions	Asset Requirements	Result
Firm resource heterogeneity Firm resource immobility	Value Rareness Imperfect imitability Non-substitutability	Sustained competitive advantage

Source: Barney (1991).

Idiosyncratic resources resulting in competitive advantage are often sophisticated technological assets and capabilities: physical technology plus the socially complex relations that enable a firm to fully exploit its physical technology add to sustainable competitive advantage (Wilkins 1989, Kotabe 1992). Based on this insight, researchers such as Hamel and Prahalad (1990) have highlighted the importance of technology by developing the 'core competence' approach (see Section II-1-1).[9]

We summarise that technological, mostly intangible, assets are recognised as being one important factor causing the (international) competitiveness of an organisation (Collis 1991). Our next argument is that because of this characteristic, technological leadership also explains a good deal of international market entry activity. This argument will be elaborated on in the following section.

---

[9]   A related concept is that of 'core technology', which is defined as the technology that gives a company its sustainable competitive edge (Wrigley 2002).

2.1.3. Technological Resources and International Market Entry

In global strategy literature, resource-based issues have been recognised as a necessary supplement to the original corporate strategy research after advances in the internal analysis of the firm and new insights on how a firm's accumulated resource stocks relate to international competitiveness (Andrews 1971, Collis 1991).

The argument is that international firms who possess (technological) assets superior to those in foreign countries can leverage these assets by transferring them to the target country, and thus can successfully compete with domestic incumbents (Hymer 1976). Monopolistic advantage can thus favour market entrants, although foreign firms generally have to overcome hurdles that do not apply to domestic incumbents.[10]

Foreign direct investment can then be viewed as the transfer of an intermediate good, called knowledge, which embodies a firm's advantage, whether it is the knowledge underlying technology, production, marketing or other activities (Kogut and Zander 1993). This link is described by Caves (1971, p. 5):

> *"Here is the link to the basis for direct investment: the successful firm producing a differentiated product controls knowledge about serving the market that can be transferred to other national markets."*

The ownership of some knowledge-based asset that provides it with a cost or quality advantage and that can be adapted and employed in multiple locations is thus a prime driver of foreign direct investment. It has even been described as the primary motivation for a firm to become multinational (Maskus 2003,

---

[10]   Hymer (1976) calls this initial disadvantage 'liability of foreignness' and refers to the effect that firms operating abroad encounter inevitable disadvantages that host country competitors do not face. Hence, liability of foreignness is a relative concept. The disadvantages have four major drivers (Zaheer 1995): spatial distance (i.e. logistics, coordination and communication), a lack of host country roots (i.e. higher learning costs), a perceived lack of host country legitimacy (i.e. higher reputation building costs) and restrictions from the home country (e.g. export constraints for high technology).

Markusen 1995). Extending this argument leads to an explanation of the multinational corporation itself. As Buckley and Casson (1976, p.45) put it, because "knowledge is a public good which is easily transmitted across national boundaries, its exploitation is logically an internal operation; thus unless comparative advantage or other factors restrict production to a single country, internationalisation of knowledge will require each firm to operate a network of plants on a worldwide basis". Kogut and Zander (1993) therefore argue that the multinational corporation arises out of its superior efficiency as an organisational vehicle for transferring technological knowledge across borders.

The asset types that have been proposed to qualify for creating a competitive advantage on an international level are manifold. Examples include patents, machinery design, factory-floor management, chemical formulas, or construction blueprints (Hymer 1976, Maskus 2003). Emphasis has also been placed on brand names and marketing know-how (Hildebrandt and Weiss 1997) and even integrated distribution channels (Anderson and Coughlan 1987).
However, only a certain type of asset qualifies for explaining foreign direct investment, and these are assets that cannot be transferred externally via the market. As Teece (1983) argues, the driving force for foreign direct investment can be identified in the existence of specific rent-yielding assets that are non-tradable for transaction-cost reasons.

Summarising the paragraphs above, we have shown how certain technological assets enable a firm to compete successfully in foreign markets. It has also been hinted at the argument that the characteristics of a technological asset can determine *how* an international firm enters a foreign market. This leads us to the next section of this chapter, which deals with different market entry forms and their relationship to international technology transfer.

2.2. The Choice for a Market Entry Form and Technology Transfer

The preceding sections have discussed the relationship between technology and international market entry in general. We now extend the topic of market entry by distinguishing between several possibilities of market entry, and how intended technology transfer can be related to them.

The next paragraphs will give an overview of market entry forms (MEFs) in general and how the choice for a market entry form relates to intended technology transfer, while research insights relating to impact of intended technology transfer on the choice of a market entry form will be presented in the paragraphs thereafter.

2.2.1. Overview of Market Entry Forms

Governance structures for foreign market entry may "run from discrete market exchange at the one extreme to centralised hierarchical organisation at the other, with myriad mixed or intermediate modes filling the range in between" (Williamson 1985, p.16). A distinction among market entry forms is important because they represent the structural context for the value adding activities of a foreign enterprise and thus influence the business processes both within and external to the firm in many ways (Williamson 1979, 1985, Fu 2005).[11]

We therefore present a short overview of the possible market entry forms and their key characteristics. The following table offers a list of possible entry forms, in this case classified according to the basic governance structure employed.

Table 5: Classification of Market Entry Forms

Export Entry Forms	Contractual Entry Forms
Indirect	Licensing
Direct agent/distributor	Franchising
Direct branch/subsidiary	Technical agreements
Other	Service contracts
	Management contracts
Investment Entry Forms	Construction/turnkey contracts
Sole venture: new establishment	Contract manufacture
Sole venture: acquisition	Co-production agreements
Joint venture: new establishment	Contractual joint ventures
Joint venture: acquisition	Other
Other	

Source: Adapted from Root (1994), p.6.

---

[11]   We use the terms 'market entry form' and 'market entry mode' interchangeably.

The distinct feature of export entry forms vis-à-vis contractual or investment entry modes is that a company's final or intermediate product is manufactured outside the target country and subsequently transferred to it. Export is thus confined to physical products. Contractual entry forms are long-term, non-equity associations between an international company and an entity in the target country that involve the transfer of technology or human skills from the former to the latter. Finally, investment entry forms have the characteristic to involve ownership by an international company of manufacturing plants or other production units in the target country (Root 1994).

As suggested earlier, our interest is focused on joint ventures (JVs), where ownership and control are shared between one or more companies. Joint ventures can be either contractual or based on equity investment, but there are many other distinctive characteristics to it. An elaborated description of (international) joint ventures will therefore follow in Section II-4. Here we briefly describe general insights regarding the alternative forms of market entry.

Market entry forms can be categorised according to the share of capital and management allocated between the home and the target country. As the figure below suggests, export entry modes generally imply little investment of capital and management resources within the target country. The major share of value creation is accomplished in the home country. On the other hand, a sole venture with production capability represents a significant investment of capital and management resources, and a major share of value adding activity is accomplished in the target country. The joint venture is among the market entry modes with the highest commitment of capital and management resources in the target country.

Figure 8: Market Entry Modes and Capital and Management Commitment

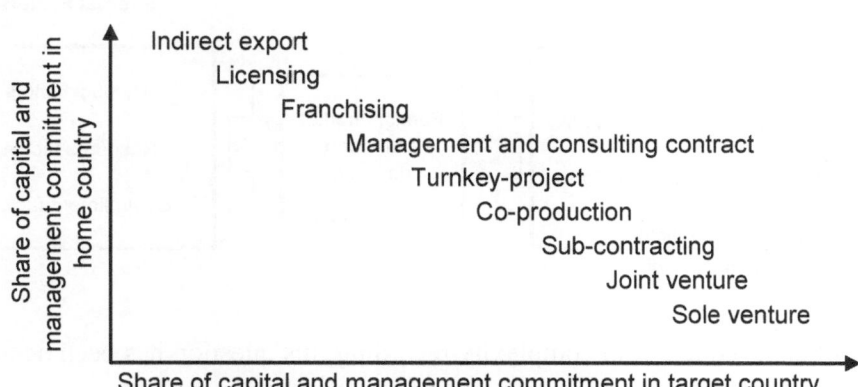

Source: Kumar (1989), p. 917.

Market entry forms with high capital commitment in the target country also tend to be those that allow – among other benefits – the transfer of sophisticated technology to the target market. This fact has been subject to intense research, because it effectively relates a firm's choice of a market entry form to the question of whether technology shall be transferred. This will be further elaborated in the following sections.

2.2.2. The Choice for a Market Entry Form

In this section, we review some literature on the choice of a firm's market entry mode. First, it is relevant to notice that the choice between entry modes depends on a wide variety of influencing factors. A detailed list of possible factors as proposed by Root (1994) includes 44 factor groups and an in-depth description of each of them would go beyond the scope of this thesis. A summary of the influence factors is shown in the figure below.

Figure 9: Summary of Factors Influencing the Entry Form Decision

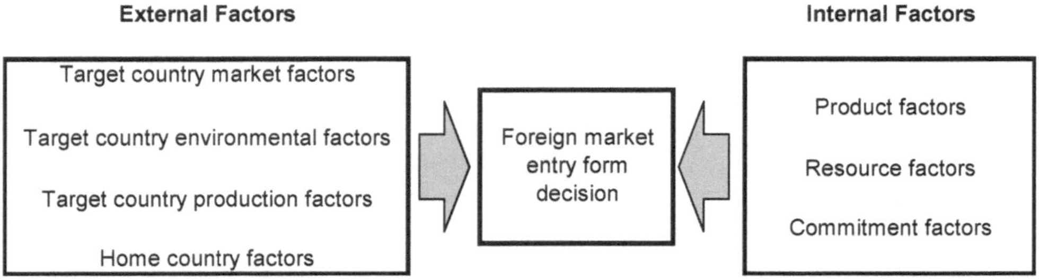

Source: Root (1994), p. 9.

A significant reduction in complexity regarding this question has been achieved by several alternative theoretical approaches. In the following paragraphs, we will first discuss the application of the transaction cost theory to the choice of a MEF. Second, we will discuss the eclectic paradigm by Dunning (1977, 1979).

Transaction Cost Theory

The credit for the first applications of the transaction cost approach in the context of international market entry can be assigned to McManus (1972), Buckley and Casson (1976), Teece (1981) and Hennart (1982). Similar to the basic considerations of transaction cost economics (Coase 1937, Williamson 1975, 1979, 1985), the above-mentioned authors investigate the factors that make firms choose internal, hierarchical governance structures to enter a foreign market rather than external, market-based solutions.[12]

Hierarchical solutions in the context of international market entry are those that involve internal coordination for the main value adding activities, such as foreign direct investments, equity joint ventures, and also direct export. On the other hand, external solutions include non-equity based cooperation forms with

---

[12] Buckley and Casson (1976) investigate the general factors influencing transaction costs and production costs in the international context and thus determine the choice of a market entry form. Teece (1981) looks at a multinational corporation's transaction costs for obtaining necessary inputs in foreign markets, such as raw materials, know-how, and capital. Hennart (1982) argues for the relevance of transaction costs related to the international transfer of goodwill and quality control.

a local partner, such as licensing or sub-contracting (Root 1994, Weiss 1996). The following table distinguishes the features of internal vs. external market entry forms.

Figure 10: Internal vs. External Transfer Channels for International Market Entry

Characteristics	External Transfer	Internal Transfer
Control mechanism	Contract	Ownership
Market entry form	Without direct investment Export Licensing Sub-contracting	With direct investment Export Sole venture Joint venture
Main value adding activities	Local contract partner	Company

Source: Adapted from Weiss (1996), p.47.

A general insight from transaction cost theory relevant to our study is that the transfer of specific assets to foreign countries would involve significant transaction costs when transferred via the market, e.g. due to negotiations, protection from possible patent infringements. Therefore, firms tend to transfer such assets within the firm boundaries – via sole ventures or joint ventures – or rely on export (Buckley and Casson 1976).

Regarding the intended transfer of technology, factors that influence transaction and production costs are the nature and sophistication of technology to be transferred to the target country, the technological sophistication of the international firm and its potential partners, as well as the countries involved (Paquin 2000).

Eclectic Paradigm

The second approach we review is the eclectic paradigm by Dunning (1979, 1980, 1993). It has been coined the 'eclectic' paradigm because three main factors identified by prior research are assembled, namely *ownership*, *location*, and *internalisation* advantages. Ownership specific (competitive) advantages allow firms to outsmart competition in foreign countries. This category of advantages is derived from industrial organisation theory and resource-based considerations, and explains why firms are able to compete abroad at all (see previous sections and Hymer 1976). Location-specific advantages relate to location theory and explain why a given activity should or should not necessarily be located in the foreign target country (Dunning 1977). Finally, internalisation advantages relate to advantages that organisations can derive from carrying out activities within the boundaries of the firm, as opposed to seeking transactions via the market. This part of the paradigm relates to the transaction cost considerations theory discussed above. The following figure shows the corresponding decision tree.

Figure 11: The Eclectic Paradigm and a Firm's Market Entry Choice

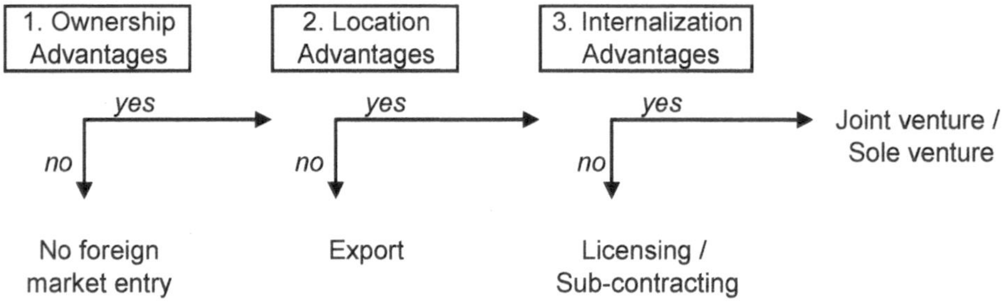

Source: adapted from Welge and Holtbrügge (2003a).

The combination of the factors can be used to derive an 'optimal' market entry form for any specific combination of firm, country, and activity.[13] Relating to joint ventures as the focus of this study, one could therefore infer that firms who enter the Chinese market by means of an international joint venture are in a position to exploit ownership advantages, location advantages, and internalisation advantages.

---

[13]   The eclectic approach explains a large deal of international firms' choice of market entry form, and is widely accepted in empirical research on direct investments. However, researchers don't agree on whether the empirical findings of the eclectic approach make up a normative model for market entry or are only descriptive (Brouthers et al. 1999).

However, the choice for a market entry form is often subject to legal restrictions, and joint ventures are in many cases formed as a consequence of such restrictions. For instance, Beamish (1985) finds in a survey that 57% of joint ventures under review were created due to government regulation. Tomlinson (1970) finds for a sample of joint ventures in India and Pakistan that the main reason for joint venture formation is government pressure. Sinha (2001) describes joint ventures as 'fall-back alternative' to sole ventures, when legal restrictions prohibit sole ventures. Root (1994) even argues that the prohibition or discouragement of sole venture entry by governments in developing countries or communist countries is the most common reason for joint venture formation.

That joint ventures often represent a second best investment entry strategy for foreign direct investors has to be taken into account when assessing the cooperative setting and the knowledge sharing behaviour among the joint venture partners. As further described in Chapter IV, we argue that partners in a joint venture formed due to government restriction might be less cooperative than partners in 'free-choice JVs', which has implications for the degree of technology transferred by the German investor and the resulting technological endowment of the joint venture.

The next section will consider the relationship between intended technology transfer and the choice for a market entry form in detail. It is important to know what technological endowment to expect, *given* that a foreign investor has decided to serve a foreign market such as China with a joint venture.

2.2.3. Technology Transfer and the Choice for an IJV

As indicated before, there are empirically proven interactions between the intended technology transfer of a market entry project and the choice of a market entry form. We here review some of the most relevant empirical studies in the respective field of research and focus on results with regard to joint ventures. This is a necessary excursus, because it sheds light on the nature of technology endowment to be expected within an international joint venture.
Empirical studies on this topic are abundant. A selection of studies that make claims about the choice of a joint venture as a vehicle for technology transfer is

depicted in the following table, together with a summary of results that relate to technology.

Table 6: Empirical Findings on Technology and the MEF Choice (Focus on IJVs)

No.	Study	Number of cases	Time period	Geographical Focus	Findings regarding Technology and Market Entry Form
1	Wilson 1977	350 TT-cases	1971	USA, World	Product innovations can better be transferred through external channels than process innovations. Complex technologies are transferred through cooperative channels.
2	Teece 1977	26 TT-projects	until 1977	USA, World	TT transaction costs decrease with firm size and transfer experience of the sender. Also with industry expertise of receiver and age of the technology. Entry forms are path dependent.
3	Mansfield et al. 1979	30 TT-cases (only one JV)	1974	USA, World	Profitable knowledge is transferred via sole subsidiaries; old or established knowledge through JVs or even externally. Process technology is transferred through sole ventures.
4	Mansfield and Romeo 1980	65 JVs	1960-1978	USA, World	The older a technology and the more often it has been transferred, the higher is the probability that firms transfer it via a JV or even externally.
5	Coughlin 1983	57 firms, 406 innovations & 548 imitations	1945-1975	USA, World, mainly EU	External channels for technology that are old and have often been transferred. Many sole ventures that are result of government regulations were endowed with old process technology.
6	Davidson and McFedridge 1984	1376 TT-cases	1945-1975	USA, World	Firms with high R&D spending and technolgy-based sales transfer technology internally. High cultural distance or entry barriers make firms opt for JVs.
7	Hladik 1988	334 TT-cases from 420 JVs	1974-1982	USA, World	JVs are chosen when similar firms pool technological competences for strategic reasons.
8	Dörrenbächer 1992	18 TT cases from 21 JVs	1991	Germany, Russia	Low technological expertise makes JVs necessary when sophisticated technology is supposed to be transferred. Imitative technologies are transferred via JVs, whereas innovative technologies are transferred via sole subsidiaries.
9	Kogut and Zander 1993	82 TT-cases	after 1960	Sweden, World	Knowledge is transferred by joint ventures, but for uncodified knowledge, the preferred vehicle is transfer between wholly owned units.
10	Niosi et al. 1995	50 Projects with 90 TT-cases	1990	Canada, LDCs	Transfer of complex technologies (i.e. a lot of training) can best be achieved through joint enterprises. High cultural distance or entry barriers lets firms opt for sole subsidiaries.
11	Katz et al. 1996	208 separate TT-cases within 1 JV	1990-1993	Germany, USA, Japan	Complex technologies with a tacit component, e.g. procedures, are transferred via JVs.
12	Kloth 1996	70 interviews with different JVs	1994-1996	Switzerland, China	High cultural distance or entry barriers reduce the choice of entry to JV vs. sole venture. Technologies with tacit component can best be transferred through a JV.
13	Weiss 1996, Hilde-brandt and Weiss 1997	78 firms with 83 TT cases.	1992-1993	Germany, World	Market entry with a technology component leads to internal transfer (sole ventures) rather than external (licensing). Marketing know-how is more decision-relevant than technology.
14	Fu 2005	52 companies (12 joint ventures)	2001	Germany, China	The degree of innovation of a technology (asset specifity) and the threat of a patent infringement (coordination requirement) increase market commitment (JV or even sole venture).

Source: Paquin (2000), p. 215. Adapted and extended.

The majority of study results point in the same direction: If firms want to transfer sophisticated technology to a foreign market, they tend to use hierarchical market entry forms such as a joint venture or even a sole venture. This occurs because these market entry forms enable firms to transfer sophisticated or tacit knowledge such as process technologies (Mansfield et al. 1979, Coughlin 1983) and procedures (Katz 1996). On the other hand, old and more established technologies that possibly have been transferred to foreign locations before (Teece 1977, Mansfield and Romeo 1980) and that are well-codifiable (Kogut and Zander 1993) lend themselves to external technology transfer such as licensing.

Within internal transfer modes, the joint venture and the sole venture are the main alternatives for the transfer of complex technologies. However, very profitable or sensible technologies have been found to be transferred through sole ventures only because joint ventures carry the risk of knowledge drain to the partner (Mansfield et al. 1979, Niosi et al. 1995). On the other hand, joint ventures are chosen when partners are either very complementary in nature (Hladik 1988), or when cultural barriers or government regulation make such an entry form unavoidable (Fu 2005).

One should also keep in mind the earlier finding (see previous section) that joint ventures are very often the result of market entry restriction and represent a second-best alternative to a sole venture. Entry restrictions have been found to influence investors' decisions for the market entry in China to a great extent (Fu 2005).

To summarise the above discussion, we can state that an international firm's intended technology transfer to a foreign market has been found to impact the choice for an entry mode. International joint ventures tend to 'host' rather advanced technological knowledge and they allow market entrants to transfer tacit knowledge. The only entry form that suits the transfer of advanced and tacit knowledge even better is the wholly owned subsidiary. However, this market entry form has other disadvantages or may be forbidden by the host government. These insights will further be discussed in next section, where we will elaborate on the international joint venture in detail, focusing on the motivations and mechanisms for knowledge sharing.

3. The International Joint Venture as a Channel for Technology Transfer

The market entry form international joint venture (IJV) is at the heart of this thesis. We therefore dedicate this section to literature on the international joint venture, providing definitions, typologies, and research regarding the typical roles and motivations of JV partners.

We then proceed by reviewing research regarding technology transfer within international joint ventures. Evidence specifically relating to partner motivations, partner roles, and knowledge absorption for international joint ventures in China will conclude this section.

3.1. The (International) Joint Venture

This section will describe the (international) joint venture in detail. We start out by providing several definitions for joint ventures. We then discuss the reasons why IJVs are formed, some joint venture classifications, and the (dis-)advantages of using joint ventures as a market entry form.

3.1.1. Definitions

Generally, joint ventures fall into the class of *alliances*, which are defined as any voluntary cooperative agreement between firms that involves the exchange, sharing, or co-development of contributions by partners such as capital, technology, or other firm-specific assets (Gulati and Singh 1998, Harrigan 1986, Parkhe 1993).

Within alliances, one can distinguish between contractual co-operations (including contractual joint ventures), minority investments, and equity joint ventures (Gulati and Singh 1998). The term 'joint venture' is thus used for both a co-operative contractual relationship and firm co-operations involving equity investment. Although both interpretations can be subsumed under the term 'joint venture', management literature generally distinguishes between the equity and the non-equity type.

A contractual joint venture (CJV) is a cooperation on the basis of a contract in which partners do not create a separate organisational entity (Hennart 1988). These contractual partnerships are predominantly used for the implementation of temporary projects (Paquin 2000), and due to the fact that no equity investment is undertaken, some authors argue that contractual joint ventures should not be considered as joint ventures in a narrow sense (Ubele 1991, Macharzina 1994).

The specific property of a joint venture in the narrow sense, the equity joint venture (EJV), is that it combines resources from more than one organisation to create a new organisational entity that is distinct from its parents (Inkpen and Beamish 1997). Partners hold equity positions in what can be regarded as a quasi-autonomous organisational entity, a 'legal child' (Yan 2000, p.4).

Many authors have identified and summarised the *defining properties* of an equity joint venture. For the sake of brevity, we only refer to one definition as offered by Guttermann (1997). Further reference is offered by Yan (2000) or Büchel et al. (1998).

As outlined by Gutterman (1997, p.7) the essential elements of an EJV as a matter of law are as follows:
1. EJVs are established by contract, expressed or implied, and consist of one or more agreements involving two or more persons or organisations which are entered into for a specific business purpose. EJVs are formed for a specific and definable business objective and are established for a limited duration, primarily because (a) the complementary production activities involve only a subset of the assets of the participants (b) the complementary assets have only limited service life and/or (c) the complementary production activities will be of limited efficacy.
2. Each of the participants to an EJV contributes property, cash or other assets and organisational capital held by them for the pursuit of a common and specific business purpose. As opposed to a contractual relationship, the contributions in the context of an EJV are made to a newly-formed business enterprise, which usually takes the form of either a corporation

or a partnership, created to pursue the business goals and objectives of the venture. As such, the participants acquire a joint property interest in the assets and subject matter of the joint venture.

3. The participants share a common expectation regarding the nature and amount of the expected financial and intangible goals and objectives of the EJV. The goals and objectives tend to be narrowly focused, recognising that the assets deployed by each of the participants represent only a portion of the overall resource base.

4. The participants share in the specific and identifiable financial and intangible profits and losses of the EJV, as well as certain elements of the management and control of it.

As described before, joint ventures are frequently used as an entry form to access international markets. If, in addition to the specifications above, the 'parent organisations' originate from different countries, the denotation is usually changed to international joint venture (IJV). This implies that the joint venture is located in the target market of a foreign investor firm, and that at least one partner is a firm domestic to the target market.[14] The term 'international joint venture' usually relates to *equity* joint ventures, not *contractual* ones.

3.1.2. Strategic Objectives for International Joint Ventures Formation

A variety of strategic objectives have been suggested to explain a foreign investor's commitment to form international alliances in general and international joint ventures in particular.

Common arguments for the formation of international alliances are the necessity gain access to local networks (Contractor and Lorange, 1988, Kogut 1988b, Hagedoorn 1993, Bennett et al. 2001), locally controlled resources (Hennart 1991) or to bridge a cultural distance gap (Root 1994). They also offer the

---

[14] Sometimes Joint Ventures are formed by three or four partners, but only in rare cases. A partnership of two firms that divide the management responsibility is the usual case (Paquin 2000 and Macharzina 1994). Partnerships of 5 or more partners are extremely rare (Zielke 1992, p.33).

advantage of saving resources compared to green- or brownfield investment (Inkpen and Beamish 1997).

Several systematic approaches have been offered to summarise the strategic objectives for alliance formation. Gulati and Singh (1998) sum up all arguments with the following eight basic rationales: (1) sharing costs/risks, (2) access to financial resources, (3) sharing complementary technology, (4) reducing the time span of innovation, (5) joint development of new technology, (6) access to new markets, (7) access to new products, and (8) sharing production facilities.

Harrigan (1986, p.16) provides a detailed overview of the goals and motives for the formation of international joint ventures. Drawing a basic distinction between internal, competitive and strategic reasons, he lists the following arguments:

Internal reasons
1. Spreading costs and risks
2. Safeguarding resources which cannot be obtained via the market
3. Improving access to financial resources
4. Benefits of economies of scale and advantages of size
5. Access to new technologies and customers
6. Access to innovative managerial practices
7. Encouraging entrepreneurial employees

Competitive goals
1. Influencing the structural evolution of the industry
2. Pre-empting competitors
3. Defensive response to blurring industry boundaries and globalisation
4. Creation of stronger competitive units

Strategic goals
1. Creation and exploitation of synergies
2. Transfer of technologies and skills
3. Diversification

These rationales are not to be regarded as mutually exclusive, however. For example, an alliance that is set up in order to share production facilities can have the beneficial effect of sharing costs, accessing a partner's financial resources, and access new markets at the same time.

3.1.3. IJV Classifications and Allocation of Partner Roles

Closely relating to the above-mentioned strategic objectives, researchers have proposed a multitude of approaches for the classification of (international) joint ventures.

As argued by Büchel et al. (1998), a joint venture classification should not only related to the functional responsibilities, but also to the relationship between the partners and the strategic focus of the joint venture. One commonly used framework that satisfies this condition has been proposed by Hermann (1988, p. 56). He distinguishes among the following six types of joint ventures:

1. *Sales joint ventures*: the producer and a local partner cooperate in an arrangement which is a mixture of independent representation and own branch. The function usually is to sell the foreign partner's product in the local partner's country or region.

2. *Market technology joint ventures*: combination of the market knowledge of one partner with the production of product know-how of the other

3. *Complementary technology joint ventures*: the partners combine their technologies to diversify their existing product/market portfolios.

4. *R&D joint ventures*: the aim is to create synergy by making joint use of research facilities, exploiting opportunities to specialise and standardise, combining know-how and sharing risks.

5. *Supply joint ventures*: competitors with similar input needs cooperate to safeguard supplies, reduce procurement costs or prevent the entry of new competitors

6. *Concentration joint ventures*: competing firms cooperate to form larger and more economical units.

For the description of international joint ventures in particular, Fu (2005) and Welge and Holtbrügge (2003a) have proposed adaptations of the framework above. The resulting types are presented in the following table.

Table 7: Types of International Joint Ventures

Type	Input Partner A (Foreign)	Input Partner B (Local)	Alliance/ JV
Sales/ Product-to-Market	Products or services	Market access to target market	Regional sales for products or services
Technology-to-Market	Technological know-how for development and production of product or service	Market access to target market	Development, production, marketing, sales
Complementary-Technologies	Technological know-how in field A	Technological know-how in field B	Combined know-how, e.g. for the development of new products
Production	Demand for products to be produced in target market	Host of production activities	Production
Sourcing	Demand for inputs to be acquired from target market	Access to local input markets	Sourcing
Concentration	Unit of organisation covering activity A	Unit of organisation covering activity A	Combination of both organisation units for the sake of concentration

Source: Adapted from Fu (2005), p. 140.

The international joint ventures investigated in the context of international market entry are predominantly *product-to-market* or *technology-to-market* JVs: the principal objective of the foreign partner is to expand its market access or tap into new markets and does so through an alliance with a partner that has marketing and distribution expertise in those markets (Inkpen and Beamish 1997, Yan and Gray 1994, Parkhe 1991).

Using the classification of Thompson (1967) between pooled, sequential, and reciprocal interaction (see Section II-1), Gulati and Singh (1988) argue that international joint ventures as used for market entry are typical examples for alliances characterised by *sequential* interaction, because the output of one partner is handed off to the other, to whom it is an input. In product-to-market joint ventures, the foreign partner brings in existing products or services to be sold in the target market. The local partner builds on these products and takes charge of local distribution and sales. Technology-to-market JVs exhibit more interaction among partners because existing technology has to be adapted to produce suitable output for the target market, but the division is also sequential: The foreign partner brings in the technology to produce the output, whereas the local partner assures that this output reaches the local customer.

Local knowledge is regarded as particularly important for a foreign firm aiming to establish an operational presence (Stopford and Wells 1972), and to overcome the 'foreign liability' associated with the fact that the market entrant is not a domestic actor (see Hymer 1976 and previous discussion on the topic). Inkpen and Beamish (1997) give an example for the importance of local knowledge provided by the local joint venture partner. They describe that when Kentucky Fried Chicken entered China, a local partner was considered essential because of the complexities associated with obtaining operating licences and leases, negotiating employment contracts, and interpreting investment regulations. The local partner also contributed important knowledge about local cultural traditions.

The following table summarises some commonly stated advantages and disadvantages of the market entry form IJV from the perspective of the foreign investor. Benefits of joint ventures thus generally relate to the possibility to create synergies (Kutschker and Schmidt 2002), e.g. by using a local partner's specific capabilities, to pool resources. Disadvantages generally stem from the fact that the local partner has his own agenda for profit-maximisation, which might or might not result in goal congruence with the foreign investor.

Table 8: Advantages and Disadvantages of Joint Ventures

Advantages	Disadvantages
Ease of access to foreign market as a sales and sourcing market	Difficulty of the assessment of potential partners
Ability to use production capacity in the target market inspite of limited resources	Goal incongruence with the partner regarding the JV's conception
The local market knowledge of the partner saves investment in acquiring such knowledge	Insufficient degree of control
Possibility to build a national identity together with the partner	Integration and communication problems
Ease of management of relations to national government and other institutions	Risk of uncontrolled knowledge drain to the partner
Size economies and pooling of resources	

Source: adapted from Fu (2005, p.142) and Freericks (1998, p.132).

Partnering with a local partner can thus provide international firms with a low-cost, fast access to new markets by 'borrowing' a partner's local infrastructure which is already in place (Doz, Prahalad and Hamel, 1990). This infrastructure includes sales forces, local plants, market intelligence, and the marketing presence necessary to understand and serve local markets.

However, IJVs also entail certain drawbacks. A commonly noted disadvantage of using IJVs is the risk of *uncontrolled knowledge drain* to the partner. This will be discussed in the following section, as part of a general discussion on knowledge exchange and absorption within international joint ventures.

3.2. Technology Transfer within International Joint Ventures

One necessary condition of a successful joint venture is the productive interaction between the partners. This productive interaction relies on knowledge flows and inter-partner learning. However, while firms benefit from teaming up with a joint venture partner that possesses relevant complementary resources or in order to share knowledge, a common intention is to acquire some of the partner's knowledge or resources over time. This might even lead to "a race to learn", where the partner that learns the fastest benefits most from the relationship (Hamel 1991). Mutual learning between partners then leads to a paradoxical situation, because partners sacrifice the uniqueness of their contribution and thus their value to the co-operation, especially in the case of a public good such as technology.

In the following section, we discuss how an international joint venture enables the transfer of sophisticated knowledge between partners and how it can be susceptible to the misuse of knowledge contributions.

There exists a large body of both theoretical (Kogut 1988, Westney 1988, Parkhe 1991, Pucik 1991) and empirical (Hamel 1991, Dodgson 1993, Simonin and Helleloid 1993, Inkpen 1995, Inkpen and Crossan 1995) work on knowledge exchange within joint ventures. A general paradox that is described is that on the one hand, IJVs might be chosen as a market entry form to potentially allow for – or even enhance – inter-organisational knowledge sharing. On the other hand, firms that engage in alliances can find themselves competing on the basis of their strategic resources, with knowledge being an increasingly important one (Loebecke et al. 1998).

Irrespective of the partners' intention to encourage, tolerate, or prevent knowledge exchange within joint ventures, a certain degree of knowledge exchange is inherent in a joint venture relationship. Inkpen and Beamish (1997) argue that organisational boundaries become permeable when a JV is created. This permeability provides firms with a "window on their partner's broad capabilities" (Hamel et al. 1989, Hamel 1991). Even embedded knowledge of other organisations and, therefore, new organisational skills and capabilities can be learned. This process, which Huber (1991) calls "grafting", allows firms to internalise a partner's knowledge not previously available within their own organisation.

Furthermore, the acquired knowledge is not only constrained within the boundaries of the joint venture. Through processes of amplification and interpretation, acquired knowledge has the potential to be shared and distributed within the parent organisation (Daft and Weicke 1984, Nonaka 1994). Eventually, the acquisition of a partner's knowledge may decrease a firm's dependency on its partner and thus lead the joint venture to be dissolved unilaterally (Inkpen and Beamish 1997).

As a result, knowledge exchange among IJV partners is a double-edged sword. While some knowledge exchange is sought after by both parties in order to create synergies, each firm faces the risk of harmful appropriation of its own proprietary knowledge by the partner. Previous researchers have examined concerns of moral hazards and appropriation of skills and knowledge in alliances (Pisano et al. 1988, Pisano 1989). It is found that the risk of such appropriation is exacerbated if an alliance involves technology sharing (Inkpen and Beamish 1997). This is especially the case when the limits of the technology to be exchanged are difficult to specify (Merges and Nelson 1990, Annand and Khanna 1997, Oxley 1997).

Because sophisticated technology is usually an input of the foreign JV partner, foreign firms have concerns about the appropriation of their skills by their local partner (Oxley 1997).[15] As a result, firms that employ easily appropriable

---

[15]   More specifically, firms are concerned to capture a fair share of the rents an alliance generates. This concern arises due to the combination of behavioural uncertainty, difficulties to specify intellectual property rights, and the challenges of contractual monitoring and enforcement (Teece 1986, Levin et al. 1987)

technology and want to serve a foreign market with that technology are even likely to refrain from working with a local partner because of appropriation concerns (Gulati and Singh 1998, p. 788).[16]

However, appropriation concerns can also be of concern for the local partner. The key contribution of the local partner usually is the knowledge of the local environment, which is thus the key source of the foreign partner's dependence on the partnership (Yan and Gray 1994). As a foreign partner increases its knowledge of the local market, the foreign partner's need to rely on the local partner dissipates (Inkpen and Beamish 1997). The unique domain of the local partner then shifts from being complementary to the foreign partner to being undistinguished (Ring and van de Ven 1994).

Empirical evidence shows that actual appropriation of knowledge occurs by both foreign and local partners. Hennart et al. (1995) report the behaviour of Japanese JVs in the United States, who were likely to terminate their JV agreements with local partners after having acquired the necessary local knowledge to run the business on themselves. Ono (1991) describes the cases of Ralston Purina, Bayer AG, Monsanto Co., and Sandoz in Japan, who quit international alliances and turned their activities into wholly owned subsidiaries after having acquired the necessary local knowledge.

Contrary to that, researchers consider the acquisition of advanced technical knowledge by the local partner rather challenging (Inkpen and Beamish 1997). Hamel (1991) finds that market intelligence is generally transferred between alliance partners more easily than knowledge of leading-edge manufacturing skills. It is argued that technical skills, such as complex engineering processes, are too much embedded in organisational routines (Nelson and Winter 1982) to be easily extracted by the local JV partner. Furthermore, the technical skills provided by a foreign partner will often consist of a complementary combination

---

[16]   As defined by Teece (1986), the appropriability of one firm's technology depends on both the inherent replicability of a certain firm-specific knowledge and the strength of the intellectual property regime protecting it. Gulati and Singh (1998) study international alliances in three worldwide industries over a 20-year period. They find that appropriation concerns are the most significant factor influencing the chosen governance structure of an alliance.

of both explicit and tacit knowledge (Nonaka and Takeuchi 1995), and only the tacit dimension will enable the proper use of the explicit knowledge.

For the case of China, the 'typical' division or partner roles has been described by Yan and Gray (1994, p. 1492). In their study of U.S.-Chinese joint ventures, they refer to a U.S. manager saying: "We have the technology and certain know-how. The Chinese partner knows how to make things happen in China. You put the two things together right, it works".

More specific insights can be gained by some recent survey results among Chinese and foreign partners to international alliances (Kasperk et al. 2006). As the following table shows, foreign partners to international alliances are predominantly interested in gaining access to the Chinese market, its distribution networks, and local human resources. On the other hand, Chinese partners are predominantly seeking access to product and process technologies when engaging in an alliance.

Figure 12: Goals of Chinese and Foreign Partners in International Alliances

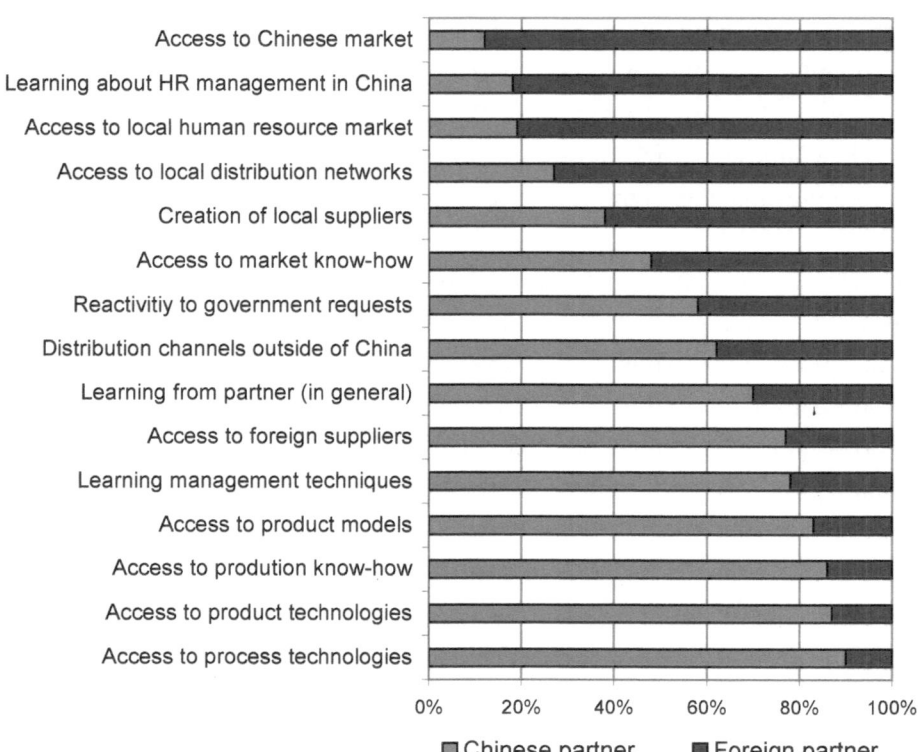

Source: Kasperk et al. (2006), p.181.

Business literature quotes many examples where the appropriation of a foreign partner's knowledge has occurred without its consent (Kasperk 2006, Dyer and Maier 2007). However, Yan and Gray (1994) report that local joint venture partners in China encounter significant difficulty in learning from their foreign partners. This is because foreign partners frequently take explicit measures to protect the transparency of their skills, particularly if the skills comprise explicit knowledge held by a few experts (Inkpen and Beamish 1997).

We can thus summarise that both a foreign and a local partner of a joint venture puts proprietary knowledge at the risk of potential diffusion when endowing the commonly-owned joint venture with it. From the foreign investor's point of view, this is mainly related to proprietary product and process technologies. The next section will provide a detailed insight of the target market: China.

4. China as a Target Market for International Market Entry

In this section, we describe the recent history of China regarding political and economic developments. We will also provide an overview of the current regulatory framework regarding foreign market entry, intellectual property protection, and technology transfer. We end this chapter with a review of empirical data regarding technology transfer to China.

The history of the Chinese civilisation can be traced back for more than five thousand years. Historians distinguish between 25 dynasties until the 20th century (Fu 2005). Looking at the last two thousand years of world history, China was the world's leading nation with respect to political, economic, economic, and technological strength for most of this time (Lin 2000). Europe overtook China as late as the 18th century as a result of the industrial revolution (Fu 2005). China, on the other hand, was hit by economic crisis, revolts, and an explosive population growth at the beginning of the 19th century, transforming the country into one of the world's poorest. Nowadays sees the 'rise of the dragon' at an astonishing speed – and many authors predict that China's GDP will overtake the GDP of the USA as soon as 2020 (Holtbrügge and Puck 2005).

4.1. The Political and Economic Development in China since 1978

The following paragraphs will describe the development of the economic and political environment since 1978. Our review of recent economic and political developments starts with China's transition from a planned economy to a market economy, following the decision to open up China to the world economy. After the death of Mao Zedong in September 1976, this transformation was agreed on at the Third Plenary Session of the 11th Congress of the Communist Party of China (CPC), in December 1978. With reforms such as the 'Four Modernisations' programme as well as the opening up of the Chinese economy to outside investors, the new leadership under Deng Xiaoping started a process that is still ongoing today (Fu 2005).

Yue (1997) has described China's transformation since 1978 and its main goals as an evolutionary process that can be divided into four phases (see figure 13).

Figure 13: Transition of the Chinese Economic System since 1978

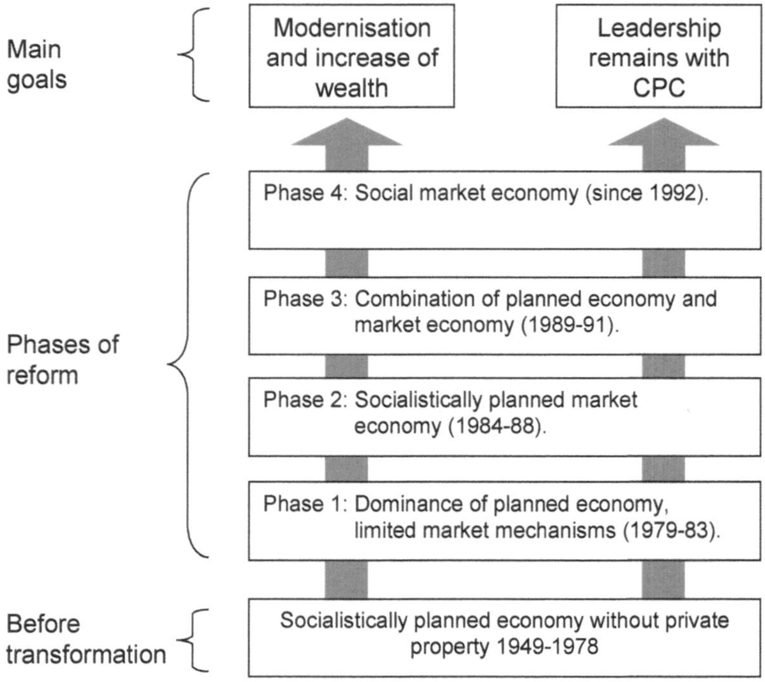

Source: Adapted from Yue (1997) p.69.

These four phases can be described as follows (see Yue 1997) unless stated otherwise):

- Phase 1: Although the system still relied on a planned economy, limited sectors and products were opened for market price mechanisms. This applied to the agricultural sector and was later extended to the financial sector in general and banks in particular. In 1980, the special economic zones Shenzhen, Zhuhai, Shantou and Xiamen in the provinces Guangdong and Fujian were founded. These were subject to preferential investment conditions and were supposed to attract foreign direct investment (Fu 2005).

- Phase 2: The reforms were extended to the industrial sector and were combined with a step-wise liberalisation of the economic system. Also, in 1984, the foreign trade system was reformed. Up until then, Chinese foreign trade was exclusively processed by 12 state-owned organisations. After the reform, the provinces and later individual companies received the right to decide on the import and export of certain goods themselves (Holtbrügge and Puck 2005). The market pricing mechanism now applied to a whole series of industries and the Chinese currency. In 1985, some existing special economic zones and 14 coastal cities were defined as 'open economic zones', and additional preferential economic zones were established (Fu 2005).[17]

- Phase 3: Because of accelerating inflation and frictions within the system, the focus between 1989 and 1991 was on the achievement of equilibrium and the consolidation of the economy.

- Phase 4: At the 14th Plenary Session of the Communist Party, the social market economy was set as a target for China. Although macroeconomic conditions were still to be under government control, the allocation of resources was to be determined by demand and supply. The former secretary general of the CPC Jiang Zemin then proclaimed (Holtbrügge and Puck 2005, p.7):

  "The objective of the reform of the economic structure will be to establish a socialistic market economy that will further liberate and expand the productive force [...]. In this way, we shall provide an incentive for

---

[17]   We do not discuss special economic zones and their development in detail. We refer the interested reader to Holtbrügge and Puck (2005) and Fu (2005).

enterprises to improve their performance, so that the efficient ones will prosper and the inefficient ones will be eliminated."

As Fu (2005) argues, the process after 1978 in China was a *gradual* transformation from a planned economy towards a market economy. Or, as Clifford et al. (1997, p.41) put it, "China is, slowly, and fitfully, becoming a market economy".

China's reforms were accompanied by unprecedented economic growth. Between 1978 and 2005, the GDP per capita increased from 381 RMB to 14.040 RMB. The development of the both the GDP and the GDP per capita depicted in the following table.

Figure 14: China's Gross Domestic Product between 1978 and 2005

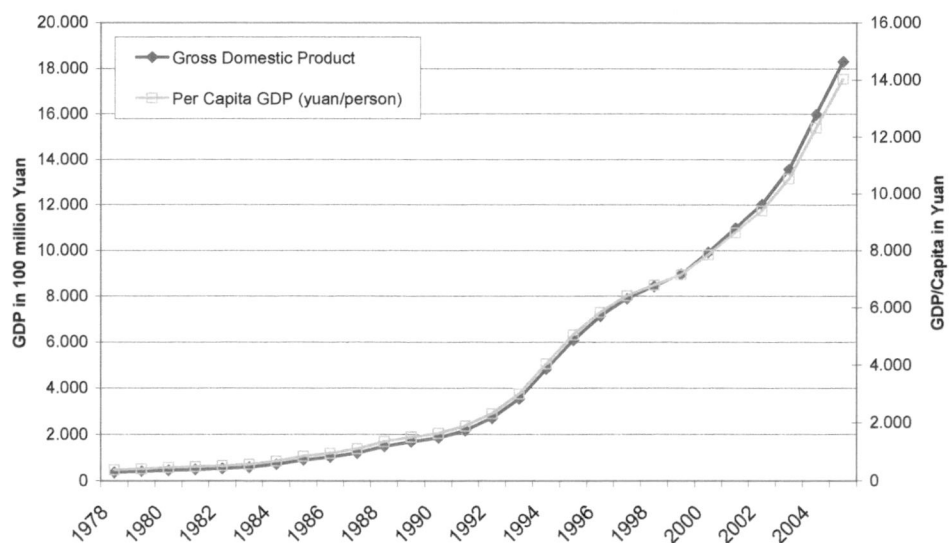

Source: National Bureau of Statistics of China.

A crucial element of the modernisation was China's opening to the outside world and the corresponding growth of international trade and foreign direct investment. As Bennett et al. (2001) put it, the influx of foreign capital resulting from international trade has been among the major factors driving China's growth in recent years.

China also joined the most relevant international multilateral organisations and agreements, such as the United Nations, the WTO, the World Bank, the International Monetary Fund, the Asian Development Bank (ADB) and the ASEAN. In addition, China has entered a large number of bilateral treaties encouraging trade and investment (Fu 2005).

The results of China's foreign economic policy regarding international trade and foreign direct investments are considerable. Between 1978 and 2005, exports increased from USD 9.8 billion to USD 762 billion, whereas imports increased from USD 10.9 billion to USD 660 billion. Also, exports have consistently exceeded imports in the last ten years, and the Chinese current account surplus exceeded USD 100 billion in 2005 alone.

Figure 15: Chinese Imports and Exports between 1978 and 2005

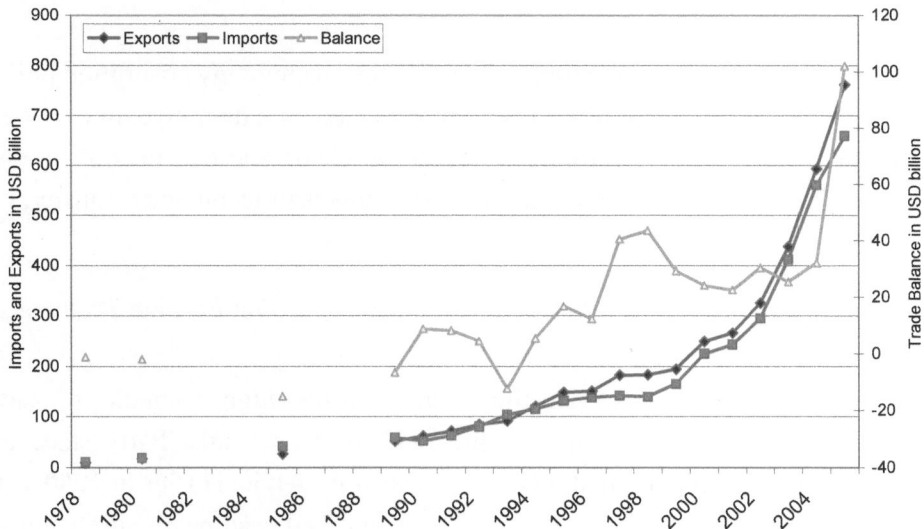

Source: National Bureau of Statistics of China.

Summarising the discussion above, it is apparent that the economic reforms and the corresponding decision to open up China to the world economy have contributed to a period of unprecedented economic growth and international trade activity.

Of special importance in the economic development and international trade policy of the Chinese government has been the accumulation of modern technology, as the next paragraphs will describe.

4.2. Chinese Policies and Regulations regarding Technology Transfer

Since 1978, the Chinese government has aimed to modernise the Chinese economy, ultimately in order to increase the wealth of its people. One of the government's measures to achieve this goal was the policy to accumulate modern technology and capital by means of technology transfer from abroad. This policy has influenced the regulatory framework for international trade and foreign investment (Konrad 1989). Technology transfer to China is thus not only affected by Chinese government policy, but is also a long-term goal for it. Before we discuss the regulations affecting market entry to China by foreign firms in Section II-4-3, we therefore describe the Chinese policy regarding technology transfer.

We start out by describing the roots of the Chinese technology absorption policy and the measures by which it has been implemented. We then give an overview of the intellectual property protection regime in China and conflicts of interest that stem from knowledge drain via informal channels of technology transfer.

4.2.1. General Policy of the Chinese Government regarding Foreign Technology

Song (2004, p.66) argues that the "magic words high technology" have dominated the thinking of Chinese elites and the Communist Party since the reform in 1978. He reports that a book published by Alvin Toffler in 1980 with the title 'The third wave' had been declared required reading for the Chinese leadership cadre. In his book, Toffler (1980) develops the hypothesis that with the help of high technology, developing countries like China can catch up with leading industrial nations by skipping the phase of industrialisation, leading them directly from the agricultural society to the information age (Seitz 2002).

Since then, modernisation by means of technology absorption remains one overarching goal of Chinese policymaking after the opening of the Chinese economy.[18]

A major milestone was the 'Four Modernisations Programme', encompassing Science and Technology, Agriculture, Defence, and Industry. This programme placed particular emphasis on transfer of technology from advanced western economies and was implemented using a number of incentive schemes for foreign investors, technologically advanced companies, and export-oriented companies (Steinbach 2004).

For example, the State Council promulgated the *Provisions of the State Council of the People's Republic of China for the Encouragement of Foreign Investment* in October 1986.[19] These provisions granted international joint ventures a number of privileges including preferential taxation, simpler licensing procedures, the freedom to import input materials and equipment, more autonomy from bureaucratic interference, interest free loans, and the right to retain and swap foreign exchange (Bennett et al. 2001).

The logic of the 'deal' between foreign investors and China as it continues to be practiced today has been described as follows: investments and technology transfer in return for the possibility to service the attractive Chinese market (Gersemann et al. 2002). The more a company contributes to Chinese goals, the more welcome it is (Langhauser 2000, Shi 1994). Song (2004) argues that the Chinese government offers a dilemma to the foreign investor – an entry ticket to the Chinese market in return for technology transfer.

---

[18] Of course, the import of technology was already a goal of the Chinese government prior to 1978. As described by Bennett et al. (2001), the main source of technology imports were turnkey projects, mainly in heavy industries such as steel, machinery, and vehicle manufacture. In the 1950s and early 1960s, the main trading partner was the USSR. After the cooling of relations with the USSR in the late 1950s, Japan, Western Europe, and the USA gained a significant portion in such turnkey projects, mainly in petrochemical, steel, electricity generation equipment and mining machinery industries. However, a review of technology transfer policy in 1978 found that the turnkey approach was deficient in a number of respects: they it was expensive and provided limited scope for developing Chinese technological capability (Bennett et al. 1997).

[19] All laws and regulations mentioned in this text are listed in the reference list together with a recently accessed official Internet source.

In the next few paragraphs, we will describe both the 'rules of the game' for foreign investors in China and how technology transfer is encouraged by them. Additionally, we give examples of how the Chinese government takes an active role in acquiring technology based on these rules.

4.2.2. Regulatory Framework and the Government's Role

In the current Chinese regulatory framework, technology transfer by foreign investors is generally addressed by two sets of regulations. On the one hand, it is addressed by regulations governing foreign trade. On the other hand, it is subject to China's industrial policy regulations that include rules regarding foreign direct investments.

Technology transfer through *foreign trade* is addressed by §15, §17, §18 and §19 of the *Foreign Trade Law of the People's Republic of China*. These paragraphs form the basis of the *Regulations on Technology Import and Export Administration of the People's Republic of China* that have been passed in May 1985 and last revised in January 2002 in accordance with WTO regulations.

These regulations govern the import and export of technology, from defining the overall goals of the Chinese policies and the administrative system to be implemented down to such operative regulations as the procedural guidelines for the approval of technology import applications.

Generally, the regulations encourage the import of advanced and applicable technologies (Art. 7). In particular, foreign partners in technology import contracts (also including licensing agreements) must guarantee that the technology provided is accurate, effective and capable of achieving the agreed technical object (Article 25). Also, certain restrictive clauses on such contracts that limit the use of technology to the importer are prohibited (Article 29).

The other main legal tool to govern technology transfer is China's *industrial policy* and the corresponding regulations on foreign direct investments. The most relevant legal sources here are the *Provisions of the State Council of the People's Republic of China for the Encouragement of Foreign Investment* of

October 1986, the *Provisions on Guiding Direction of Foreign Investment* and the *Catalogue for the Guidance of Foreign Investment.*[20]

The *Provisions on Guiding Direction of Foreign Investment* and the *Catalogue for the Guidance of Foreign Investment* distinguish among prohibited, restricted, encouraged and permitted investment projects. This categorisation mainly applies on an industry level, but can also refer to specific current government projects.

Foreign enterprises are excluded from 'prohibited' industries. Typical examples are media and industries relevant for the national security (Holtbrügge and Puck 2005, p.66). Foreign enterprises may become active in 'restricted' industry sectors, but only after the approval of strict regulations and/or subject to quotas. For 'encouraged' industry sectors, the Chinese government offers specific investment incentives.

The categorisation of an industry as 'encouraged' is derived from the socio-economic demand of PR China (Fu 2005). The main factor that makes a sector fall under this category is the Chinese government's desire to attract advanced technology and to make the Chinese industry independent from – or competitive in – global markets.

Examples of qualification criteria that relate to the import of advanced technology as mentioned in Article 5 of the *Provisions on Guiding Direction of Foreign Investment* are:

- Projects employing new high technology and advanced practical technology which can improve performance of products, increase efficiency of enterprises, or manufacture new equipment or new materials while domestic productivity is deficient; and
- Projects adopting new technology or equipment that can conserve energy and raw materials, utilise resources and renewable resources in a comprehensive way or prevent environmental pollution.

---

[20] The *Interim Provisions on Guiding the Direction of Foreign Investment* were approved by the State Council on June 7, 1995, and have been replaced by the *Provisions on Guiding Direction of Foreign Investment* on April 1, 2002. The *Catalogue for the Guidance of Foreign Investment Industries* is regularly revised and published by the Ministry of Commerce.

Furthermore, Article 8 of the *Provisions on Guiding Direction of Foreign Investment* states that the *Catalogue for the Guidance of Foreign Investment* may determine that a foreign invested project can be carried out "only in the form of equity joint venture or contractual joint venture", "with the Chinese party holding the majority of shares", or "with the Chinese party holding the relative majority of shares". This especially applies for projects in restricted industries.

Although the discussion above could be extended to cover more detail, we restrict ourselves to the most relevant legislation and refer the interested reader to sources such as Fu (2005) or Huck (2005).

Instead, we provide some evidence on how the Chinese government *actively* participates in technology negotiations. Besides setting a regulatory framework as described above, the Chinese government is known for taking an active role in maximising the technological inflow from abroad. The following two examples illustrate this point.

One well-known example is the first Chinese automotive joint venture, the Shanghai Volkswagen Automotive Company Ltd., which was established after intense technology transfer negotiations with potential investors. As Dong (2004) describes, the Chinese ministry of industry had invited numerous leading automobile manufacturers such as GM (USA), Nissan (Japan), Toyota (Japan), Renault (France), Citroen (France), Volkswagen (Germany) and Mercedes (Germany) to take part in a bid for the partnership in a joint venture to service the Chinese market on an exclusive basis. The choice of Volkswagen was the result of its commitment to fulfil all of the ministry's requirements, especially regarding the degree of technology transfer. Other bidders, such as GM, had rejected the offer by the Chinese government.

In another case – again involving the automobile industry – political pressure from the Chinese government was a key decision factor for GM's decision to invest in an R&D centre in Shanghai (Gassmann and Zhen 2004). In the late 1990s, there was intense competition between several global automotive companies concerning the establishment of an automobile joint venture in Shanghai, which was speculated to be the last approval of this kind for many years. General Motors finally won the licence.

70

We can summarise this section with the observation that Chinese legislation is very much geared towards the attraction of foreign technology and thus provides the basis for the technology acquisition efforts of Chinese companies. Huck (2005) even argues that *all* Chinese economic policy regulations vis-à-vis international companies consequently aim at the acquisition of capital, modern technologies and management know-how.

4.2.3. Knowledge Drain and the IP Protection Regime in China

From the point of view of investors, China is frequently categorised as a risky target for foreign market entry due to a perceived lack of IP protection and – correspondingly – the risk of knowledge drain following technology transfer (Geissbauer 1996, World Economic Forum 1995). In this section, we cite some evidence for knowledge drain in China, before we describe the evolution of China's IP protection regime until now.

Knowledge Drain in China

China is one of the world's largest markets for product imitations, and for some Chinese regions, e.g. Zhejiang, Jiangsu and Anhui, the copying of Western brands represents the most important economic sector (Kasperk et al. 2006). In 1995, almost 20.000 product imitations were registered, 3.000 of which were affecting the products of foreign firms (Fu 2005). Kasperk et al. (2006) estimate that in 2003, about 15–20% of Western brands and 50% of replacement parts sold in China were imitations.

Knowledge drain also occurs within international joint ventures, as the following examples illustrate. A prominent case for imitation has been the recent developments for Volkswagen in China. Volkswagen, a pioneer in the Chinese market, has had very successful years with its German–Chinese joint ventures until the recent past. However, since 2004 domestic competitive pressure has significantly affected VW's business and VW's market share fell to around 15% in 2006 (Kasperk et al. 2006). One reason has been the emergence of the Chinese car manufacturer Chery in 2004. Apart from being extremely popular among the Chinese people and having been elected the 'official car for congress representatives' in the same year, the Chery mainly shocked the VW management because 60% of its parts were identical imitations of VW-Jetta parts (Gärtner 2004). Another recent example is ABB, the Swiss power and automation technologies giant. As Gassmann and Zhen (2004) describe, ABB has lost plenty of technological knowledge through their Swiss-Chinese joint-ventures, mainly because of high turnover rates of personnel. The final example involves the food manufacturer Danone and its Chinese joint venture partner Wahaha. After several years of cooperation, Wahaha apparently built several production sites for food products parallel to the Danone-Wahaha joint venture, selling almost identical products to the originals produced by the joint venture (Dyer and Maier 2007).

How can the large product imitation numbers and the examples of product imitation be explained? On the one hand, the sense of guilt for copying ideas is lacking. Imitators are often proud of their ability to manufacture products of high quality. The Chinese saying goes: "A good copy honours the master" (Trempel 2001, p. 35).

On the other hand, because the imitation industry contributes significantly to the economy of certain regions, it exerts influence on local administrators. As argued by Song (2004), many provincial governments are backing local manufacturers and do not shy away from confronting foreign firms. According to the Chinese saying "The sky is high, the emperor is far away", the national government's influence ends at the 'entrance to the village', so that policies for the protection of intellectual property by the national government are hard to implement on the provincial level (Trempel 2001, p.13). As argued by Kasperk et al. (2006), piracy firms have even been reported forming alliances to increase efficiency and product quality. For end consumers, 'good' imitations are becoming almost indistinguishable from the original products.

However, there are several developments in China that exert pressure for a better technology protection regime. First, pressure for an improvement of IP protection been constantly exercised by Western (especially U.S.) firms and recently also by innovative national firms (Fu 2005, Kögel and Gälli 1997). Second, China has evolved into one of the world's leading investors in R&D

(Song 2004). Finally, China's membership to the WTO in 2001 entailed the implementation of IP related regulations. The evolution of China's IP regime has therefore been subject to a lot of improvement, as the following paragraph will argue.

Evolution of the Chinese IP Regime

The practice of legal protection of intellectual property is not new in China, although it has never been given much emphasis (University of Göttingen 2001). The first patent law was released in March 1984 and put into force one year later (Fu 2005). As late as 1985, the first national copyright office was founded (Song 2004) and China joined the International Convention for the Protection of Industrial Property ('Paris Convention') (Kasperk et al. 2006, Birden 1998). In the same year, China and Germany ratified a bilateral agreement for the protection of foreign invested capital that also addressed the protection of intangible assets (Huck 2005).

The accession of China to the WTO on December 11th, after 15 years of intense negotiations, has had a significant impact on regulations concerning technology transfer. As a consequence of WTO membership, China now acknowledges the following five multilateral agreements as binding (Huck 2005, p.2):
- GATT 1994 (General Agreement on Tariffs and Trade)
- GATS (General Agreement on Trade in Services)
- TRIPS (Trade Related Aspects of Intellectual Property Rights)
- DSU (Dispute Settlement Understanding)
- TRPM (Trade Policy Review Mechanism)

Since then, the Chinese regulatory system has been in a constant process of adaptation, and some of the pre-WTO laws governing foreign activities have been revised: according to the Chinese constitution, international legislation overrides national law, so that national law had to be either completed or replaced (Kasperk 2006). Accordingly, the national patent law was reformed after 2001 and a number of new regulations have been passed.

A multitude of laws and regulations has been passed that relate to the protection of Intellectual Property, as shown in **Table 9**. The table shows the names of laws and regulations as well as the effective date and the date of the latest amendment.

Table 9: Laws and Regulations regarding IP Protection in China

Laws, Regulations or Rules	Effective (E) and Last Amendment (L) Date
Trademark Law of the People's Republic of China	E: Mar 1st, 1983 L: 23.05.2007
Patent Law of the People's Republic of China	E: Apr 1st, 1985 L: Aug 25th, 2000
Copyright Law of the People's Republic of China	E: Jun 1st, 1991 L: Oct 27th, 2001
Rules for Pesticide Administration	E: May 8th, 1997 L: Nov 29th, 2001
Regulations on the Protection on New Varieties of Plants	E: Oct 1st, 1997
Regulations on the Protection of Layout-Design of Integrated Circuits	E: Oct 1st, 2001
Implementing Regulations on Patent Law	E: Jul 1st, 2001 L: 23.05.2007
Regulations on Computer Software Protection	E: Jan 1st, 2002
Management Regulations on Audio and Video Products	E: Feb 1st, 2002
Regulations on the Protection of the Olympic Symbols	E: Apr 1st, 2002
Implementing Regulations on the Copyright Law	E: Sep 15th, 2002
Implementing Regulations on the Trademark Law	E: Sep 15th, 2002
Regulations for the Implementation of Drug Administration Law	E: Sep 15th, 2002
Regulations on the Customs Protection of Intellectual Property	E: Mar 1st, 2004
Regulations on Administration of Veterinary Drugs	E: Nov 1st, 2004
Interpretations by the Supreme People's Court and the Supreme People's Procuratorate on Several Issues of Concrete Application in Handling Criminal Cases of Infringing Intellectual Property	E: Dec 22nd, 2004
Regulations on the Copyright Collective Administration	E: Mar 1st, 2005
Implementation Rules for the Regulations regarding the Protection of New Varieties of Plants (Agriculture part)	E: Jun 16th, 1999
Implementation Rules for the Regulations regarding the Protection of New Varieties of Plants (Forestry part)	E: Aug 10th, 1999
Implementation Rules for the Regulations on Integrated Circuit Design Protection	E: Oct 1st, 2001
Management Measures on Wholesale, Retail and Rent of Audiovisual Production	E: Apr 10th, 2002
Management Measures of Audiovisual Production Import	E: Jun 1st, 2002
Provisions for Identification and Protection of Well-known Trademarks	E: Jun 1st, 2003
Procedure for the Registration and Administration of Collective Marks and Certification Marks	E: Jun 1st, 2003
Measures on Patent Agency Administration	E: Jul 15th, 2003
Measures for the Enforcement of Copyright Administrative Penalty	E: Sep 1st, 2003
Measures for the Implementation of Regulations governing Customs Protection and Intellectual Property Right	E: Jul 1st, 2004

Source: MOFCOM (2004).

As the table shows, most important regulations were revised after 2001, resulting in an IP protection regime that is now theoretically achieving the standards as required by WTO members.

However, the practical implementation of these laws is not always guaranteed. In spite of high sanctions and strong efforts for the improvement of IP protection, there is still a gap between legal theory and practice (Fu 2005). Also, there are large regional differences regarding the implementation of the laws. As Zürl (2002) describes, while some coastal regions and cities have developed into modern, knowledge-based economies, major parts of China are still underdeveloped and employ primitive technologies with low productivity. Steffens (2004) argues that because of this difference, one has to differentiate between IP protection regimes: the coastal regions that tend to be more developed offer comparably high protection for intellectual property, while this is considerably less so in the rest of the country.

Firms that are affected by a violation of their IP rights can theoretically use legal means to prosecute the imitators. However, authors such as Ballhaus (2005) and Kasperk et al. (2006) recommend facing imitators directly and negotiating a solution, because this approach usually leads to results faster.

We summarise that the government's efforts in recent years have significantly improved the environment for foreign investors with regard to IP protection. However, due to the problems of implementing these efforts, a foreign firm about to engage in an IJV partnership in China must be considered as rational when taking a significant risk of knowledge drain into account.

The next section will provide legal details on the existing Chinese legal framework for allowed market entry forms and their implications for technology transfer.

4.3. Legal Provisions for Relevant Market Entry Forms

The following paragraphs will give an overview of the legally allowed forms of foreign market entry. Where applicable, we also mention implications for technology transfer by foreign investors.

We focus on joint ventures and their regulations regarding technology transfer. For the sake of completeness, we also give short explanations of the legal provision for the market entry forms 'representative office', 'licensing' and 'wholly foreign-owned enterprise' (WFOEs).[21] Next to the regulatory aspects of each entry form, we will focus on the implications for the technology transfer to local partners.

4.3.1. Representative Office and Licensing

Since 1980, Chinese law allows the creation of representative offices of foreign enterprises (Fu 2005). Representative offices are not categorised as foreign direct investments, because capital transfers are only allowed to a very limited extent and contracts for profit-oriented transactions may not be signed (Hilger 2001, Kutschker 1997). They are meant to represent the foreign enterprise: to create and keep up contacts – especially with the Chinese authorities, to conduct market analyses, or to collect other kind of information (Sommer 2001, Hilger 2001). Practically, however, representative offices are used for the full range of marketing and distribution tasks necessary to support business activities of foreign firms (Kutschker 1997).

The formation of a representative office is related to a very limited transfer of the entrant's technological knowledge to Chinese partners. Individual employees working for the representative office may have significant knowledge about the entrant's technological base, but there is little or no systematic transfer of technology involved.

---

[21] In the Chinese context, the previously described sole venture is called 'wholly foreign-owned enterprise' or WFOE. We will from now on use these to terms interchangeably.

Licensing agreements transfer commercial IP rights (for example trademarks, patents, or copyrights) or also non-protected know-how (e.g. process or management know-how) from one company to another. In exchange, the receiving party – the licensee – usually pays a licence fee to the originating party – the licensor (Kasperk et al. 2006). Commonly, the transfer of these intangible assets is accompanied by technical services to ensure the proper use of the assets.[22] In China, licensing agreements are categorised as technology import contracts (Hilger 2001), and are thus governed by the *Regulations on Administration of Technology Import and Export* as well as the accompanying implementing regulations.

Technology transfer to local partners can reach very high levels in the case of international licensing and the degree of local investments and management control by the foreign partner is comparatively low (see Section II-2-2) Foreign licensors must therefore rely on the functioning of the local IP protection system. Licensing is often quoted as the market entry form with the highest risk of losing control over one's proprietary technology and thus of creating additional competitors (Root 1994), especially for China (Kasperk et al. 2006).

4.3.2. Foreign Invested Enterprises

According to §22 of the *Regulations on Technology Import and Export Administration*, technology may be used as a form of investment capital in the formation of a foreign invested enterprise comprising:

- Wholly foreign-owned enterprises (WFOEs),
- Contractual joint ventures (CJVs), or
- Equity joint ventures (EJVs)

The regulatory framework concerning these three forms of investment is mainly given by the *Provisions of the State Council of the People's Republic of China on Guiding Direction of Foreign Investment*. Here, we will present legal details regarding these forms of foreign direct investment, first discussing *wholly*

---

[22]   The core of a licensing agreement – the transfer of intangible assets related to technology – distinguishes it from other contractual arrangements, such as management and technical assistance agreements. These only provide professional services to the recipient firm (Root 1994).

foreign-owned enterprises (WFOEs), then *contractual joint ventures* (CJVs) and then – finally and most extensively – *equity joint ventures* (EJVs).

Wholly Foreign-Owned Enterprises (WFOEs)

WFOEs are legally allowed since the *Wholly Foreign-Owned Enterprise Law of the People's Republic of China* (WFOE-law) was issued in 1986. However, they have only become a real option for foreign investors since the issue of the *Company Law* in 1994 and the *Preliminary Catalogue for the Guidance of Foreign Investment Industries* in June 1995.

Because the Chinese government favoured joint ventures over WFOEs for a long time (mainly to guarantee technological competence and management know-how through the foreign partner), the regulations and restrictions for the creation of WFOEs are comparatively tough (Fu 2005). According to the WFOE-law, sole ventures are only acceptable if the activities of the foreign firm contribute to the development of the Chinese economy, promise to be economically profitable, and either:

- use modern technologies, machines, and equipment that help save resources (especially energy), improve existing products, develop new products, replace existing imports or
- derive at least 50% of their turnover from exports and the amount of foreign currency received exceeds the amount of currency dispensed.

In the revised version of the WFOE-law of 2000, the constraint to derive at least 50% of turnover from exports was dropped (Fu 2005). Also, the possibility of the acquisition of Chinese local companies has been created (Kasperk et al. 2006).

WFOEs are among the most popular market entry forms in recent years, mainly because of the foreign companies' reluctance to cede control and technological know-how to Chinese partners (Kasperk et al. 2006).

Contractual Joint Ventures

Contractual or cooperative joint ventures (CJVs) comprise all forms of co-operation between foreign and domestic firms for short-term projects, hotel projects or investment deals (Holtbrügge and Puck 2005).

Contractual joint ventures are managed by a Board of Directors (BoD) or a Joint Management Committee. The distribution of seats within this board may be decided upon by the partners (Kasperk et al. 2006). In cases where a CJV entails the formation of a separate legal entity, the relation between equity capital and debt is governed by official regulations in order to ensure the entity's viability. If no separate legal entity is created, the invested assets remain the property of the cooperating partners, and profit or losses accrue to the partners according to a contractually defined division scheme (Chu 1987).

A special form of CJVs is the category 'BOT-projects' (build-operate-transfer), in which foreign companies build turnkey facilities for the use of Chinese clients. This kind of cooperation is usually used for infrastructure projects, but may also be applied for contract manufacturing, assembly or refinement processes. As Fu (2005) points out, BOT projects allow Chinese firms the acquisition of foreign technology and know-how without having to dispense much foreign currency to buy the technology involved.

Equity Joint Ventures

Equity joint ventures (EJVs) represent the oldest form of foreign investment entry. The *Equity Joint Venture Law* (EJV-Law) was issued in 1979, and the *Implementing Regulations* stem from 1982. The last revised version of the EJV-Law was issued in March 2001 (Huck 2005).

An equity joint venture is a separate legal and organisational entity located in China, which is commonly owned by a foreign investor and a local partner. Liability by the investing partners is limited to the registered investment capital, making this legal company form comparable to the English 'Ltd.' or the German

'GmbH'. The presence of a Chinese legal or natural person as one of the joint venture partners is required.

The governing organ of an EJV is the Board of Directors (BoD). This board has to consist of at least three members: one representative of each partner and the president (CEO); this CEO is the legal director of the enterprise and has to be Chinese. Effectively, however, management control may be in the hands of the foreign partner because the CEO function is mainly representative.

The minimum share of total invested capital by the foreign partner is 25%. Dependent on the industry, there also exist minimum capital share requirements as a function of the total invested capital (Kasperk et al. 2006). Registered capital may be brought in the form of cash or contributions in kind. The profits of the EJV are distributed among the partners according to their shares in registered capital.

As described by Fu (2005), Chinese companies engage in equity joint ventures with foreign firms primarily to increase their competitiveness regarding western technology and management know-how, and to encourage the transfer of capital and foreign exchange. In typical EJV setups, the Chinese partner thus contributes land utilisation rights as well as existing buildings and facilities, while the foreign partner provides cash, production equipment, and technological know-how. The contribution of foreign technology as capital investment is permissible, as long as the contributed technology is advanced and really serves the interests of China.

Regarding the contribution of technology as investment capital, the *Implementation Regulation to the EJV-law* provides for the following conditions:

- Machines and production facilities must be necessary for the production (§24).
- The appendix to the EJV contract must contain comprehensive technological documentation including copies of the patent documents and documents regarding technical characteristics (§26).[23]

---

[23] Non-patented technologies are admissible as capital contributions in Contractual Joint Ventures (§8 of the *CJV-Law*), but not in EJVs.

- Approval by the respective governmental institution (§ 27).
- Implementation of the foreign partner's contribution within the contractually mentioned timeline (§28).
- The technology must be practically applicable and advanced (§41). [24]
- Products have to contribute significantly to Chinese society and the economy or must be competitive on world markets (§41).
- Approval of the technology transfer contracts regarding the following aspects (§43)
  - Fair and reasonable costs for using the technology.
  - No restriction regarding regions, quantities, and prices of exportable goods;
  - The governing time period may exceed 10 years only in exceptions.
  - After 10 years, the importing party may keep using the respective technology.
  - The conditions for mutual exchange concerning the upgrading of the respective technology have to be equal.
  - Prohibition of restrictive clauses that are prohibited by Chinese laws and regulations.

As Fu (2005) describes, equity joint ventures are approved for a period of 10 to 30 years, but may be prolonged without much difficulty. For the liquidation of an EJV, the consent of all founding partners as well as the directors of the original registration office is necessary.

As this section has shown, there exist a variety of detailed regulations that foreign firms have to respect when investing in China. These are especially detailed with regard to technology transfer. Out of all market entry forms described here, the joint venture provides for the highest exposure of a foreign investor's technology due to detailed documentation requirements, and the approval process and required concessions to be made to the Chinese partner. Close interaction with a local firm then enables knowledge exchange between partners (as previously discussed in Section II-3-2).

---

[24]   If the foreign partner contributes old technology and that leads to damage for the local partner, the foreign partner is required to compensate for the damage (Huck 2005).

In the next section, an overview of empirical data regarding technology transfer to China is provided.

4.3. Technology Transfer and FDI to China from the World and Germany

The following paragraphs will give a brief empirical overview of the technology transfer activity from foreign countries to China. The next section will evaluate investment activity between the world and China, whereas the following section will investigate the activity between Germany and China. This is only a brief discussion, so we refer the interested reader to Dougherty (1997), who also provides detailed information regarding data sources and measurement issues for technology transfer activity to China.

4.3.1. Technology Transfer to China

Two relevant measures commonly employed to represent technology transfer include the value of high technology imports and the utilised value of foreign direct investment (Dougherty 1997). The development of these two indicators between 1978 and 2005 are shown in the figure below.

Figure 16: Technology Flows to China between 1978 and 2005

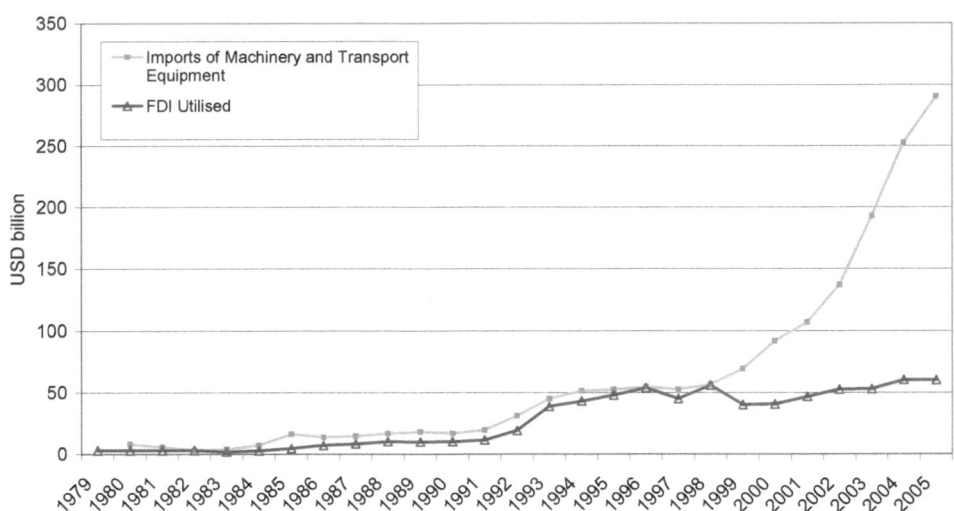

Source: National Bureau of Statistics of China.

From this data, the general increase of technology flows and the relative importance of each source are evident. Up to 1998, the import of machines and the value of foreign direct investments (FDI) account for an approximately equal value and were both growing. However, the value of imported machinery really picked up after 1999 and exceeded USD 290 billion in 2005.[25]

---

[25]   Dougherty (1997) points out that import figures for capital goods are an imperfect measure of technology transferred, as licensing and inter-company contacts are not included. Selected publications by MOFCOM refer the technology imports as the sum of imported machinery and technology transfer agreements such as license contracts (MOFCOM 2005a), so technology transfer agreements should be taken into account as well when assessing technology transfer. Unfortunately there are no time series data available for such technology transfer agreements for the whole period between 1979 and 2005, because this data is only registered by the Ministry of Foreign Trade (MOFTEC) but could not be accessed by the author.

Within foreign direct investments, the use of specific entry forms has experienced significant changes since 1979. The development of the relative shares for each FDI type is shown below.[26]

Figure 17: Foreign Direct Investments by Entry Form between 1979 and 2001

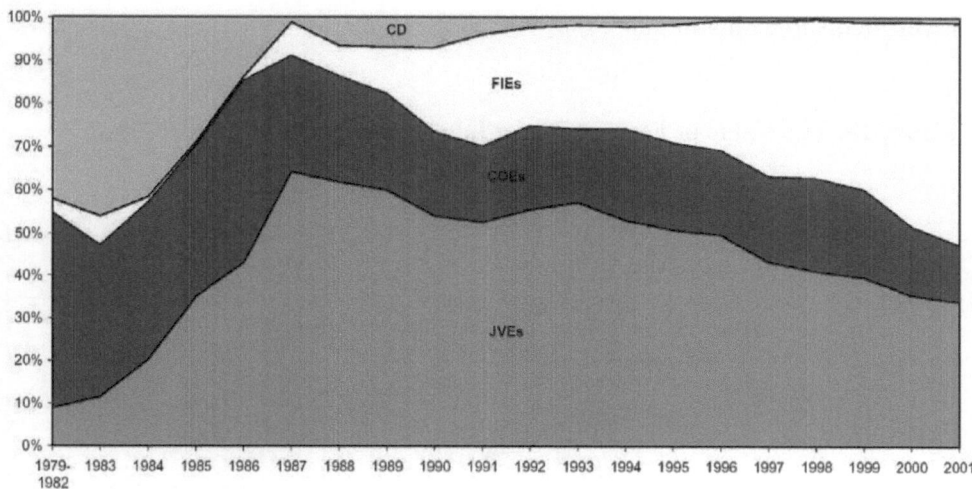

Source: Greeven (2004 p. 16). Original source: China Statistical Yearbook 2002.

As Greeven (2004) points out, cooperative operation enterprises and cooperation developments were almost exclusively used as FDI entry forms until 1992. Their relative importance declined when more structural investments were allowed and foreign investors made stronger commitments.[27]

Joint ventures have represented the dominant market entry form since 1987, accounting for more than half of all investments made. The relative importance of joint ventures then steadily decreased after 1993 as sole ventures were gaining importance.

---

[26]   Joint ventures are labelled as 'JVEs', cooperative operation enterprises are labelled as 'COEs', WFOEs are labelled as 'FIEs' and cooperation developments are labelled as 'CD'.

[27]   This is linked to the regulatory reforms regarding foreign firms, such as the amendments to the EJV Law (1990) and the *Income Tax Law for Enterprises with Foreign Capital and Foreign Enterprises* (1991).

As the figure below shows for the most recent data available, the *absolute* value of utilised FDI for joint ventures has stabilised around USD 15 bn since 1999, while the total value of FDI has increased to over USD 60 bn in 2005. This expansion is mainly due to the increased investments in WFOEs, but apparently not at the expense of investments in joint ventures. In 2005, investments in international joint ventures represented over 24% of all foreign direct investments to China in 2005.

Figure 18: The Value of Foreign Direct Investment to China 1998-2005

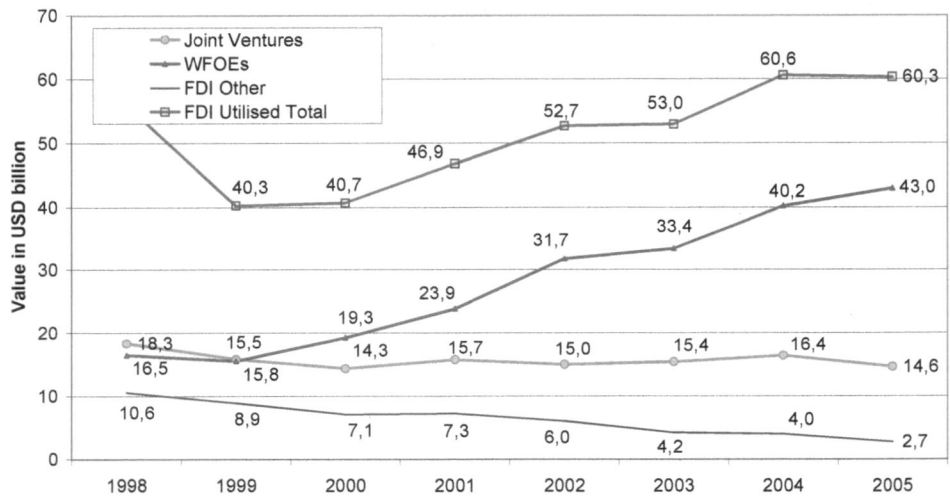

Source: *National Bureau of Statistics of China.*

One can therefore summarise that while the significance of sole ventures as a source of technology transfer to China has expanded, the market entry form international joint venture retains crucial importance.

The next section will present some specific data on the technology transfer activity from Germany to China, focusing on figures on foreign direct investment.

4.3.2. FDI from Germany to China

This section provides the reader with empirical data on the FDI activity from Germany to China. As pointed out in Chapter I, Germany is among the leading sources of foreign direct investment in China. The following figure (reproduced from figure 2 for illustrative purposes) shows that in 2005 German direct investments accrued to USD 1.5 bn. This makes Germany the second-biggest western investor after the USA.[28]

Figure 19: Top 10 Sources of FDI to China in 2005 (in USD million)

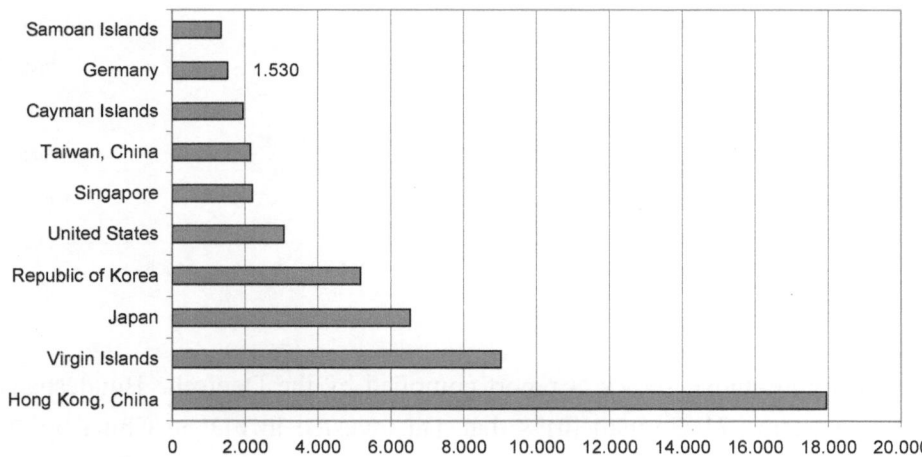

Source: National Bureau of Statistics of China.

As reported by Holtbrügge and Puck (2005), the capital stock of German FDI in China increased by a factor of ten between 1995 and 2003, reaching EUR 7.9 bn in 2003. They further estimate that about two-thirds of German investments stem from industrial firms, especially the automobile industry, the electronic industry, the chemical industry and the machine tool industry.

---

[28]   One has to take into account that some investments coming from Hong Kong and the Virgin Islands probably originate from German investors.

The following figure shows the most recent development of the stock of German FDI in China as well as the number of German firms that maintain subsidiaries in China. Both figures have risen considerably in recent years.

Figure 20: German Stock of FDI in China and Number of German Firms in China

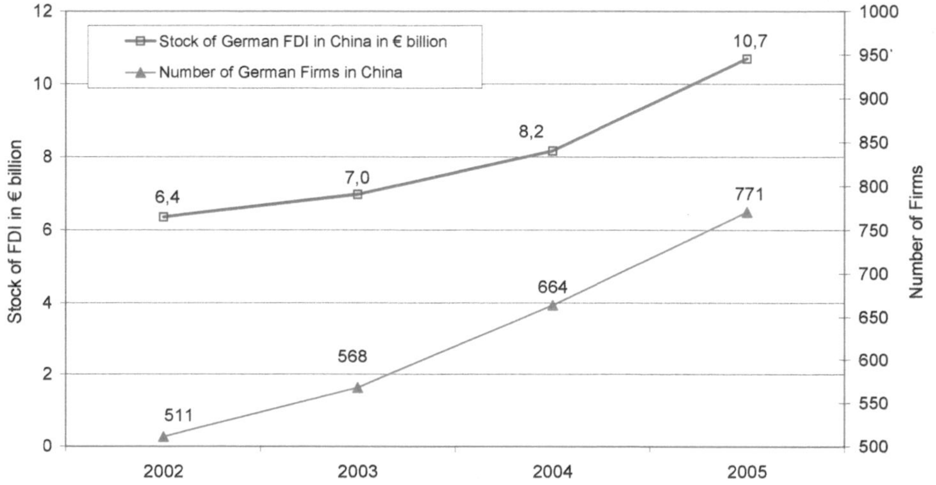

Source: Deutsche Bundesbank (2007).

As the figure above shows, a report compiled by the Deutsche Bundesbank in 2007 registers 771 German firms that had invested in FDI in China by 2005. This information can be complemented by the database of the Delegation of German Industry (AHK China) in Beijing that collects 'self-administered' registrations by German firms. In June 2006, the database contained 484 equity joint ventures, 1069 WFOEs and 32 contractual joint ventures.[29]

We can summarise our findings in this section by saying that the amount of German investment in China is considerable and that the importance of international joint ventures is quite considerable: The number of joint ventures is almost half as high as the number of WFOEs in 2006.

---

[29]   The difference between these two data sources might stem from the fact that the data from Deutsche Bundesbank only includes consolidated data, whereas the AHK database often registers several entities owned by the same company group.

The next section will provide the reader with a recapitulation of the literature reviewed in Chapter II with reference to our research questions. Chapter III will then proceed by presenting the theoretical framework of this thesis.

5. Recapitulation

As a recapitulation, we now briefly summarise the main findings of Chapter II that are relevant for the further argumentation of this thesis.

We argue that international joint ventures allow for the extensive sharing of technological knowledge, even if it is tacit and deeply embedded in certain processes. While the local partner usually provides local market knowledge, the foreign partner invests technological knowledge or other capabilities necessary to successfully market a product or service in the target market.

Second, partners do appropriate the other partner's knowledge. This occurs as part of the cooperative agreement and in order to create synergies. To a certain extent, however, this happens due to the competitive element remaining between any two profit-oriented actors. Both firms' knowledge pieces can be appropriated by the partner.

Third, and finally, technology sharing seems to be a key element for joint ventures in China, because Chinese firms prioritise the acquisition of technologically advanced knowledge. This is apparent from the empirical evidence quoted above and a result of a variety of circumstances, including the political and regulatory environment in China. Foreign investors therefore have to take the potential diffusion of proprietary technological knowledge into account when they decide to endow an international joint venture in China with sophisticated technology.

Resulting from the above-described circumstances, one *challenge* that foreign investors face is the appropriate endowment of foreign operations with technology, given that the market entry form equity joint venture is used. The question is: what lets a foreign firm *deliberately share* its technology with its joint venture partner as a result of co-operative considerations, given that a certain risk of knowledge misuse exists and given that one works in a co-operative market entry form? Restating the research question from Chapter I, our particular research questions are:

What factors influence the sophistication of the technological endowment that an international joint venture in China receives from its German parent? and
In what way do strategic considerations regarding inter-firm cooperation and knowledge sharing influence the foreign investor's behaviour?

In Chapter III, we review literature and selected theoretical frameworks that relate to the strategic rationales guiding knowledge sharing by joint venture partners. Insights from these frameworks are then translated into empirically testable hypotheses.

Chapter III: Theoretical Considerations and Derivation of Hypotheses

As was discussed above, we are committed to exploring the narrowly defined issue of technology transfer by German firms to international joint ventures in China, focussing on strategic considerations with regard to the cooperative behaviour and knowledge sharing between partners.

This chapter is structured as follows: Section III-1 explores theoretical approaches that can be used to predict the effect of cooperation in German-Chinese joint ventures and its impact on the degree of technology invested by the German firm. This discussion allows us to derive our main hypotheses.

Given that the degree of technological knowledge transferred to an IJV may be the result of a variety of factors such as technical feasibility, regulatory restrictions or economic forces, the leeway for adjusting this degree as a result of technology-sharing considerations can be quite limited. Section III-2 therefore sheds light on general factors as to why firms transfer technology into a foreign market.

The insights and predictions derived in these two sections are the conceptual basis for the structure of the empirical study, which will presented in Chapter IV.

1. Cooperation in IJVs and Technology Transfer from a Theoretical Perspective

A firm's choice to form a joint venture with another firm – and possibly share technology with it – is the result of strategic considerations, and therefore this choice is subject to the same profit-maximising guidelines as all other decisions by the firm. When firms decide to engage in a joint venture with another firm, they acknowledge the fact that they can best reach their goals by relying on the critical inputs of one or more other parties. Partners "need others to advance their individual interests" (Gray 1989, p.6). Once a joint venture is established,

© Springer Fachmedien Wiesbaden GmbH, part of Springer Nature 2008
M. Hoeck, *Cooperation and Technological Endowment in International Joint Ventures: German Firms in China*, Edition KWV, https://doi.org/10.1007/978-3-658-24355-5_3

firms must cooperate to ensure that the combination of critical inputs is transformed into a productive entity (Inkpen and Beamish 1997). As Loebecke et al. (1998, p. 19) put it, "the potential to create synergies stresses the potential for firms to improve their competitive position by cooperating. Synergetic value only exists if both players exchange knowledge."

However, profit-maximising behaviour of JV partners may imply non-cooperative behaviour, because each partner may find it "advantageous to maximise his own gains at the expense of the venture" (Hennart 1991, p. 486). And, as stated by Buckley and Casson (1988, p. 34), all parties involved have an "inalienable de facto right to pursue their own interests at the expense of others."

Models trying to explain under what circumstances JV partners end up cooperating and sharing knowledge within their partnerships have to take account of these dynamics. One school of thought that can usefully be drawn on here is the literature on static and dynamic games (Gibbons 1992a, Gibbons 1992b, Rieck 1993, Osborne et al. 1994, Dutta 2000), in particular game-theoretic work on knowledge sharing within alliances (Parkhe 1993, Bruck 1996, Loebecke et al. 1998).[30] [31]

The chosen starting point for our analysis is a simple prisoner's dilemma framework regarding knowledge exchange between organisations (Schrader 1990, Parkhe 1993, von Hippel 1988, 1994 and Loebecke et al. 1998) and its implication for decision-making in International Joint Ventures (building on Bruck 1996). However we also use insights derived from related strands of literature to widen the scope of the analysis.

---

[30] The distinct characteristic of game theoretic models vis-à-vis other theoretical approaches is the investigation of decision processes within specific structures or situations where rational, profit-maximising behaviour by several participating agents is taken into account (Bruck 1996). The benefit for an agent facing a decision is then not only a consequence of his own actions, but also of other agents' actions (Rieck 1993). The game theoretic approach is therefore well suited for the application on cooperation within alliances.

[31] The work by Parkhe (1993), who uses game theoretic reasoning to relate inter-firm cooperation to the structure of alliances and investment commitments, especially has provided a basis for our considerations.

Is is worthwile to point out at this point that not a single model will be discussed, but a "basket" of theoretical considerations that relate to the topic. Predictions provided by a particular approach are used rather heuristically when translated into hypotheses. This approach reflects the exploratory nature of this study and the aim to generate a broad picture of the effect of a Joint Venture's cooperative setting on the technology sharing commitments by a foreign investor.

This section is structured as follows. The next paragraphs will present a brief discussion on why actors might not cooperate even after alliances are formed. This paves the way for our considerations on what factors *do* influence cooperation in alliances. In Section III-1-1, we elaborate on the basic prisoner's dilemma paradigm (Rapoport and Chammah 1965) and its application to knowledge exchange. In Section III-1-2, we then present a framework for the analysis of international joint ventures and derive some first hypotheses. Extensions of the basic framework and the derivation of additional hypotheses will be the subject of Section III-1-3.

Why Alliance Formation does not Guarantee Post-Alliance Cooperation

A firm's benefits derived from cooperation are a function of both partners' behaviour. This creates a situation of mutual interdependence, a condition in which one party is vulnerable to another party whose behaviour is not under the control of the first (Parkhe 1993). For the prediction of cooperative behaviour and knowledge sharing within an international joint venture, the critical challenge is that although a positive expected payoff from mutual cooperation is necessary for alliance formation, it is not sufficient for post-alliance cooperation.

As Parkhe (1993) describes, the driving force behind any alliance formation is each participant's assignment of a net positive value to expected alliance outcomes. If a zero value is attached to the outcome of mutual cooperation, little incentive exists for firms to cooperate. And if mutual cooperation is likely to generate even negative outcomes, for example through the leakage of core competencies, it would be highly irrational to engage in cooperation. This implies that the emergence of cooperation between self-interested, autonomous

93

firms reflects anticipated individual gains that are only realisable as the fruit of joint action.

In some cases, the realisation of benefits through cooperation might require successful bargaining prior to the alliance formation. A symptomatic characteristic of many international joint ventures is thus the intense bargaining among partners during their establishment. As Rouach (2003, p.24) puts it, the agreement on an IJV contract includes a period of "aggressive negotiation, risk management and cross-cultural concerns." With respect to German JVs in China, Goeritz (2006) points out that technology negotiations with potential JV partners "belong to the most tenacious negotiations that a company has to conduct in China".[32]

If contracts were perfectly determined and the behaviour of the parties involved were bound by their agreement, one could argue that once the above-mentioned contract negotiations are concluded and agreement has been reached, the resulting joint venture is conflict-free. The JV partners would enter the 'implementation' phase, in which the agreed-on projects are implemented and royalties are collected (Rouach 2003, p.24).

Furthermore, the condition that mutual cooperation leads to positive profits is a necessary condition for alliance *formation*, it is not sufficient to promote *post-alliance cooperation* (Parkhe 1993, Bruck 1996). This is because the competitive element within a joint venture is carried into the relationship, and whether cooperative behaviour (and thus knowledge sharing) is optimal depends on a firm's incentive structure. In other words, the ultimate goal of firms to increase their profits may, depending on the situation, imply co-operative or non-cooperative behaviour with regard to knowledge sharing. As Brandenburger and Nalebuff (1996, p.38) point out, "what matters is not whether others win – it's a fact of life that they sometimes will – but whether you win … sometimes the best way to succeed is to let others do well, including your competitors".

---

[32]   Answer to open question catalogue concerning technology negotiations by Mr. Leif Goeritz, General Manager of the German Centre in Beijing.

Under a specified set of assumptions, German firms involved in a German-Chinese JV may thus have a positive incentive to share knowledge *because* it is the knowledge sharing that creates value. However, a different incentive structure might very well lead to the contrary outcome. These considerations pave the way for the *prisoner's dilemma paradigm* that is used as a starting point to disentangle the mechanisms that lead to cooperation.

1.1. A Simple Prisoner's Dilemma Paradigm and its Application to Knowledge Exchange

There is a large body of theoretical research (von Hippel 1988, Borys and Jemison 1989, Oliver 1990, Powell 1990, Ring and van de Ven 1992) and empirical research (Heide and Miner 1992, Osborn and Baughn 1990, Seabright et al. 1992, Scherer 1995) that addresses inter-firm cooperation. Yet, most research has stopped short of identifying the structural dimensions of alliances that may be tied to inter-firm cooperation (Parkhe 1993). This requirement has specifically been met by applications of the prisoner's dilemma framework and its extensions, which will be discussed in the following paragraphs.

The prisoner's dilemma paradigm (Rapoport and Chammah 1965) is a well-known framework of thought and has found its way into many game theoretic standard books (e.g. Gibbons 1992a, Gibbons 1992b, Osborne and Rubinstein 1994, Dutta 2000). A description of the game is reproduced from Parkhe (1993, p. 796):

> *"In this game, two players are suspected of a hypothetical major crime, such as murder. They are imprisoned and held incommunicado, so each must decide whether to cooperate or to defect, without knowing what the other will do. The authorities possess evidence to secure conviction on only a minor charge, such as illegal possession of weapons. If neither prisoner squeals, both will draw a light sentence on the minor charge; this state is called a mutual cooperation (MC) payoff. If one prisoner squeals and the other stonewalls, the squealer will go free (unilateral defection, UD), and the stonewaller, or 'sucker', will draw a*

very heavy sentence (unilateral cooperation, UC). If both squeal, both will draw a moderate sentence (mutual defection, MD).

Each prisoner's preference ordering is UD > MC > MD > UC. Each prisoner will be better off squealing than stonewalling, no matter what his partner chooses to do, because UD > MC (the temptation of cheating) and MD > UC (the fear of being cheated upon). But if both defect, both do worse than if both had cooperated (MC > MD). Hence the dilemma."

A graphical representation of this dilemma that uses the value elements mentioned above (MC, MD, UC, UD) is offered in the figure below.

Figure 21: Representation of the Basic Prisoner's Dilemma

		Prisoner A	
Prisoner B		Stonewall (=Cooperate)	Squeal (=Defect)
Stonewall (=Cooperate)		MC / MC	UD / UC
Squeal (=Defect)		UD / UC	MD / MD

Source: Adapted from Gibbons (1922a), Gibbons (1992b), Dutta (2000), Parkhe (1993).

Besides the conflict of dividing a given value, the prisoner's dilemma includes the potential of productive value creation by the players. It thus belongs to the class of 'non zero-sum' games, because the size of the combined payoffs varies according to both players' behaviour (Bruck 1996).

This basic prisoner's dilemma framework has been applied to the question of knowledge exchange of cooperation by a number of authors, such as Schrader

(1990), Parkhe (1993), von Hippel (1988, 1994), Bruck (1996), and Loebecke et al. (1998). Most models rest on the following set of assumptions (Bruck, 1996):[33]

- Knowledge pieces are complementary.
- Knowledge can be acquired at no cost if a partner shares.
- There exists a 'monopolistic' value of one's knowledge that accrues to a sole owner of a certain knowledge piece.
- Sharing leads to some synergy.
- Sharing can be seen as one-shot game, because once the knowledge is shared, it is gone.

Taking these assumptions as given, knowledge exchange can be described as a prisoner's dilemma because although knowledge sharing could create superior returns, the probability of the other party's defection and the opportunity costs of sharing (monopolistic value) can lead cooperating firms to 'stonewall' when it comes to knowledge exchange. These insights have been derived by – among others – von Hippel (1988) and Schrader (1990).

As shown in the figure below, the basic model assumes that two agents have complementary knowledge pieces denoted as 'r'. By sharing these pieces, each firm can increase its knowledge base to 2r. However, a firm's knowledge also includes a 'value added' component reflecting the value that results from having *unique* knowledge. This component is lost by knowledge sharing. If mutual cooperation occurs, each firm receives the complementary knowledge part owned by the other firm, but also sacrifices the uniqueness of its own knowledge

---

[33]   In addition to these assumptions, we also draw on four common arguments that undermine the applicability of the prisoner's dilemma paradigm to the setting of knowledge transfer among firms within alliances. 1. Contrary to the model's assumptions, the game's payoff structure is not known to the players, because the utility of cooperating with another firm is difficult to determine, let alone when the cooperation aims for results that are many years ahead. Also, the information paradox of technology (see Section II-1-1) implies that the value of a partner's technology can only be known after the sharing has taken place. However, these reductions are due to general limitations of models by greatly simplifying a 'real-world' phenomenon. A thorough discussion of the arguments can be found in Bruck (1996). Furthermore, some of the drawbacks are mitigated as a consequence of the models' extensions mentioned in the following sections.

piece. The most profitable option for both firms would thus be to receive the other agent's knowledge without having to share its own knowledge.

Figure 22: Prisoner's Dilemma Framework of Knowledge Transfer

Player B	Player A	
	Transfers knowledge	Does not transfer knowledge
Transfers knowledge	2r / 2r	2r + va / r
Does not transfer knowledge	r / 2r + va	r + va / r + va

Source: Loebecke et al. (1998), p. 17.

A prisoner's dilemma arises because whatever player B does, it is always more beneficial for player A not to share knowledge (as long as va is positive). The equilibrium strategy for this game is thus not to share knowledge. This is a 'dilemma' if the strategy of mutual knowledge would yield a higher value for both players (2r) than the equilibrium strategy of not sharing knowledge (r+va), i.e. 2r > (r+va).

To shed light on the mechanism leading to cooperation or defection in the circumstances of our study, we will next describe a model that is tailored to the situation of German firms that partner with local firms in China and possibly transfer technology to their international joint ventures.

1.2. A Prisoner's Dilemma for Technology Transfer to IJVs

We now adapt the basic prisoner's dilemma framework to the setting of knowledge exchange behaviour among partners in German-Chinese joint ventures. This will be used to specifically discuss the factors relevant for the German partner's decision to invest in an international joint venture's technological endowment.

Similar to the basic framework introduced before, we assume that both partners to the international joint venture own complementary knowledge pieces. A simplified view is that the German firm has some technological capability that could be used to serve the Chinese market with an industrial product or service: products or services get competitive due to the 'injection' of German technology.

Access to the Chinese market is only possible with local market knowledge. This knowledge is typically in the hands of the Chinese partner firm. The Chinese partner's contribution is to offer existing contacts and distribution channels to access the Chinese market. Only the combination of knowledge pieces within an international joint venture can yield profitable synergies from serving the Chinese market with a technologically demanding product or service.

Both firms *put unique knowledge at risk* in order to gain from these potential synergies. From the perspective of the German firm, sharing technology with the Chinese partner causes a loss of the technology's uniqueness and possibly its diffusion within the industry (see Section II-3-2 and II-4-2). The Chinese partner that contributes market knowledge and access to distribution channels loses some value as well: once the common joint venture operates based on established contacts, the Chinese firm cannot claim back its 'ownership' of the access to the Chinese market any more.

Each partner could thus *defect* on the common goal of the joint venture by forming the alliance and *not* investing its promised contribution. After absorbing the other partner's contribution, the cooperation could then be abandoned. As the review in Section II-4-2 shows, both foreign and local partners to international joint ventures have been observed to defect in such a way.

Whatever the exact nature of the contribution: Both partners are vulnerable to unilateral defection, although the extraordinary value of success after sharing knowledge is indisputable: incremental revenue following the successful introduction of a product to the Chinese market. In the following paragraphs, we will formalise our considerations and use them to derive some hypotheses.

1.2.1. Variable Names and Descriptions

In the table below, we list a number of possible value components that make up the benefit of knowledge sharing in IJVs. The detailed discussion and review of each factor will follow in due course.

Table 10: Variable Names for Model

Situation	Description of Value Impact	Variable
Both firms cooperate	Extra value created due to synergy	V_S
Own firm cooperates	Unilateral costs of sharing own knowledge piece	C_{AC}
Other firm cooperates	Value of other party's knowledge piece	V_{BC}
Own firm defects	Costs of own defection (image loss)	C_{AD}
Other firm defects	Costs of other party's unilateral defection	C_{BD}
All situations	Basic value of own knowledge	R_A

Source: Own computations.

Next, we assign the components to the various constellations of an adapted prisoner's dilemma, creating a specific application to knowledge sharing in international joint ventures. This figure is shown below. Because this study is mainly concerned with the technology transfer behaviour of the foreign investor (i.e. the German joint venture partner), we only take into account the respective possible pay-offs.[34]

---

[34]   Simplifying assumptions at this point are (a) that once the decision to cooperate is taken, a firm will transfer all of its unique knowledge to its partner, (b) that both firms take their decision simultaneously, and (c) that one player cannot observe the action or pay-off structure of the other player at the time he takes his own decision.

Figure 23: Prisoner's Dilemma Framework for German-Chinese IJVs

Local firm (Chinese)	Foreign firm (German)	
	Transfers technology (C)	Does not transfer technology (D)
Transfers market knowledge (C)	$R_A + V_{BC} + V_S$ $-C_{AC}$	$R_A + V_{BC} - C_{AD}$
Does not transfer market knowledge (D)	$R_A - C_{AC}$ $-C_{BD}$	$R_A - C_{AD}$

Source: Own computations. C represents 'cooperation' and D represents 'defection'

This framework is closely related to the previous one, but adds an extra value of cooperation (V_S) on top of the value of receiving the other agent's knowledge piece (V_{BC}) in the event of mutual cooperation. It also takes account of the costs of both the own firm's defection (C_{AD}) and the other agent's (C_{BD}) defection, and expresses the costs of one's own cooperation not as the loss of a positive value component, but as a negative impact on value (C_{AC}).

These alterations and extensions reflect recent literature in the field of knowledge transfer in alliances (Bruck 1996, Loebecke et al. 1998). The context-specific adaptation is relevant because the *structure of pay-offs* is widely regarded as one of the most important factors to be taken into account for the prediction of cooperative behaviour (Axelrod 1984, Oye 1986, Parkhe 1993). It is thus worth while to invest efforts in specifying the decision-framework that firms face in alliances, and specifically, international joint ventures.

The next paragraphs will describe each variable listed in Table 10 in detail:

V_S

Bruck (1996) and Loebecke et al. (1998) argue that when both firms end up cooperating, they create a *synergy value* that is defined as the extent to which cooperation yields additional value from interdependent knowledge sharing beyond the sum of the parties' individual knowledge. There is a general understanding among business practitioners and researchers of what effects are comprehended by the term 'synergy', and it is usually applied in a very broad sense (Bruck 1996). In this context, the two major sources of synergy value are arguably the creation of new capabilities (such as the ability to serve a new market with an adapted product) and efficiency gains by specialisation.

C_{AC}

Unilateral knowledge sharing has costly consequences, irrespective of the other firm's behaviour. Loebecke et al. (1998) use the term *negative reverse impact* to describe the value loss for a technology sender that occurs due to knowledge sharing. This is especially the case if firms operate in overlapping markets, because the technology receiver might use the technology sender's knowledge to improve competing products and processes. Negative reverse impact has the same logic as the term 'va' introduced in Figure 22, only that it explicitly captures a negative impact instead of the loss of a positive one.

V_{BC}

Benefits are caused by unilateral cooperation by the other party, irrespective of one's own behaviour. Loebecke et al. (1998) define the *leveragability value* as the potential of the knowledge-receiving party to increase its value by exploiting the received knowledge component 'on its own' beyond the cooperation. As Brandenburger and Nalebuff (1996) have pointed out, an alliance may have a mutual understanding to exploit this leveragability value, as access to the other party's knowledge enables both parties to benefit from additional opportunities by leverage. One should mention here that the concept of leveragability is inversely related to "transaction-specificity" of an asset as described by Williamson (1985). The less transaction-specific an asset, the higher is the receiver's potential to leverage its value.

C_{AD}

Bruck (1996) argues to take account of possible damages to a defecting firm that directly result from its own defection: companies that behave uncooperatively within alliances are subject to possible image loss. Not only can future interactions with the same partner be seriously hindered, but also with third parties that may qualify as potential partners in the future. If one firm publishes the defection of its partner, image loss spreads through the industry (Harrigan, 1986). Firms try to avoid entering an exchange with another actor who has a questionable reputation and, if avoidance is impossible, demand that the potentially opportunistic party absorbs the costs for additional contractual safeguards (Hill 1990).

In the case of German investors in China, image loss has the highest potential when the Chinese government or large, government-owned companies are involved in the joint venture. As pointed out by one of the experts interviewed by the author, defection on joint venture contracts can trigger a reaction by the Chinese government that prevents the defector from doing business in China for several years.[35]

The damage due to image loss is crucially dependent on the importance of *future* interactions with the same partner or third parties. C_{AD} thus considers effects that are – from a strictly theoretical perspective – only derivable by taking into account repetitive interaction. For example, one precondition for image loss to occur after defection is that the partner detects the non-cooperation. The behavioural transparency of players' actions thus gains importance. The shadow of the future on today's decisions is so important that we dedicate an extension of the basic framework to it (see following sections).[36]

---

[35]   Expert interview No. 6.

[36]   For the game theorist readers: one could argue that if the costs of uncooperative behaviour only stem from punishment by transaction partners in the future, the factor C_{AD} should not be mentioned at all in a one-shot static model framework. However, we think that one should distinguish between structural dimensions influencing the cost per se that can be taken into account already (such as 'having the government as the transaction partner') while other factors influencing the shadow of the future can

C_{BD}

A firm experiences damage in case its joint venture partner defects, because it has committed resources that could be used elsewhere to a non-performing project. It might even have transferred certain assets to the joint venture or have refrained from investing them in other company locations. The damage from not being able to usefully employ these assets realise as reduced sales revenue (Bruck 1996).

In the case of German-Chinese joint ventures, a German firm could effectively be 'blocked' from market entry to China if it can't establish two partnerships at the same time – e.g. due to resource constraints – and ends up with a non-performing partnership. From the perspective of the local firm, one could argue symmetrically: the Chinese firm may only cope with cooperating with one partner at the same time. If the German partner turns out to block the project, the Chinese investor may have no performing project at that point in time at all.

R_A

The resource cost component R_A is assumed to remain with the foreign firm in all outcomes. Its main purpose is to assess the cooperation as a whole when comparing it to sole ventures (WFOEs) or other market entry forms.

Bruck (1996) argues that such a value component is made up of two parts. On the one hand, *transaction costs* have been mentioned to occur in all cases of cooperation. In the case of a German-Chinese JV, these are mainly the additional costs of finding a partner, engaging in negotiations, and controlling the common operation. On the other hand, Bruck (1996) mentions costs of *inflexibility*. Because the future success of one transaction partner is affected by a changing environment, the commitment to one specific partner can hinder the establishment of another partnership at a later point in time. The inflexibility costs can thus be viewed as the opportunity costs of not being able to realise certain strategic options in the future (Gahl 1991).

---

be discussed separately (such as the length of the decision time horizon or behavioural transparency). This will be further discussed in the following sections.

For our analysis, i.e. for predicting the extent of knowledge sharing within a JV, R_A is irrelevant, because transaction costs and inflexibility costs are incurred in any scenario. These costs do not explain the result within an alliance.

Specific predictions for the effect of the pay-off structure on the knowledge exchange among firms will be derived below.

1.2.2. Basic Insights

As described by Osborne and Rubinstein (1994) and Dutta (2000), one can predict a rational player's behaviour in case a dominant strategy exists.[37] We will thus investigate the conditions under which either a) cooperation, i.e. the sharing of knowledge, or b) defection, i.e. the retention of knowledge, is a dominant strategy for the foreign firm.

For defection to occur, two inequalities must be satisfied: first, given that the local firm cooperates, it should be more profitable for the foreign firm to defect than to cooperate. This occurs if:

$$R_A + V_{BC} - C_{AD} > R_A + V_{BC} + V_S - C_{AC}$$
$$\Leftrightarrow \quad C_{AC} - C_{AD} - V_S > 0 \tag{1}$$

Second, it must also be more profitable for foreign firm to defect than to cooperate given that the local firm defects. This occurs if:

$$R_A - C_{AD} > R_A - C_{AC} - C_{BD}$$
$$\Leftrightarrow \quad C_{AC} - C_{AD} + C_{BD} > 0 \tag{2}$$

Defection is a dominant strategy when (1) and (2) both hold. Given that V_S and C_{BD} both have some constant positive value, equation (1) requires a larger

---

[37] A dominant strategy has the characteristic to be the optimal choice for an actor, irrespective of the actions of other actors (see e.g. Dutta 2000).

difference between C_{AC} and C_{AD} in order to hold. We therefore expect that if the foreign firm's costs of cooperation (e.g. loss of monopolistic value) outweigh the costs of defection (e.g. image loss), even when taking into account that potential synergies might not be achieved, the foreign firm should be expected to defect.

For cooperation to occur, profits from cooperation must outweigh the profits from defection, irrespective of the other player's action. This holds if:

$$R_A + V_{BC} - C_{AD} < R_A + V_{BC} + V_S - C_{AC}$$

$$\Leftrightarrow \qquad C_{AC} - C_{AD} - V_S < 0 \qquad\qquad (3) \text{ and}$$

$$R_A - C_{AD} < R_A - C_{AC} - C_{BD}$$

$$\Leftrightarrow \qquad C_{AC} - C_{AD} + C_{BD} < 0 \qquad\qquad (4)$$

In this case, equation (4) requires a larger difference between C_{AC} and C_{AD} in order to hold:[38] the costs of one's own defection must outweigh the costs of being defected upon, even taking into account the potential extra damage of cooperating while being defected upon (e.g. due to dependency). If (4) holds, we could therefore expect the foreign firm to cooperate.

Combining the two results above, it is possible to predict the foreign firm's behaviour for a certain set of values for C_{AC}, C_{AD}, V_S and C_{BD}: a firm engaged in an IJV is expected to share its knowledge if the costs of giving up its own knowledge are lower than the costs of defecting on the partnership plus the potential harm of being defected upon. On the other hand, the foreign firm is expected to defect if the synergy is not high enough to outweigh the difference between the costs of sharing one's knowledge and the costs of unilateral defection.[39]

---

[38]   Given that V_S and C_{BD} both have some constant positive value.

[39]   We again remind the reader that the costs of one's own defection are related to the question of how the future should be taken into account for one's current actions. The

Taking the relationship between C_{AC} and C_{AD} as a reference, this situation can be depicted in the following figure:

Figure 24: Computation of Critical Values for Knowledge Transfer

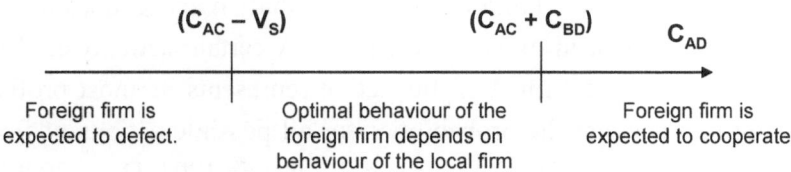

Source: Own computations.

As the figure above illustrates, it is not possible to predict the foreign firm's behaviour unilaterally if the condition $C_{AC} - V_S <= C_{AD} <= C_{AC} + C_{BD}$ holds. In this case the optimal behaviour of the foreign firm depends on the behaviour of the local firm.

If the optimal behaviour for one player depends on the behaviour of the other player, predictions can be derived by finding *equilibrium* outcomes (Nash 1950, Axelrod 1984, Osborne and Rubinstein 1994). Finding optimal strategies by finding equilibrium strategy requires assumptions about both players' pay-off structures, which is why we do not further elaborate on them.[40] Furthermore, we do not rely on a single model to derive hypothesis, but rather on a 'bundle of insights'.

Next, we complement the prisoner's dilemma framework by computing the foreign firm's optimal strategy when each of the local partner's possible actions occurs with a certain probability. We then briefly discuss the concept of inter-organisational trust.

---

influence of the shadow of the future will be elaborated in detail in one of the following sections.

[40] One common way is to assume symmetry of pay-offs. However, we do not want to rely on that assumption, because symmetry can hardly be argued to exist in the case German-Chinese joint ventures.

1.2.3. Incorporating a Local Firm's Cooperation Probability - the Relevance of Trust

This section has the aim to extend our decision analysis framework by taking into account the local firm's actions by means of cooperation probability, referring to Bruck (1996). We build on the insight that an action does not have to be unilaterally dominant in order to be chosen. A certain action can also be expected to be chosen by a firm A if this action represents its most profitable strategy, taking into account the probabilities for all possible actions of firm B (Gibbons 1992a, Gibbons 1992b, Osborne and Rubinstein 1994, Dutta 2000).

Applying this argumentation, we can compute the expected payoff of the foreign firm's cooperation as the weighted average of the payoff when matched with a cooperating local firm (with probability p_C) and the payoff when matched with a defecting local firm (with probability $(1-p_B)$). This translates into:

$$E_{foreign,C} = p_C \times (R_A + V_{BC} + V_S - C_{AC}) + (1 - p_C) \times (R_A - C_{AC} - C_{BD})$$

On the other hand, the expected payoff for the foreign firm for defection is:

$$E_{foreign,D} = p_C \times (R_A + V_{BC} - C_{AD}) + (1 - p_C) \times (R_A - C_{AD})$$

We can expect the foreign firm to cooperate if its expected profit from cooperation exceeds the expected profit from defection ($E_{foreign,C} > E_{foreign,D}$).

This relationship can be expressed as follows:

$$E_{foreign,C} > E_{foreign,D}$$

⇔
$$p_C \times (R_A + V_{BC} + V_S - C_{AC}) + (1 - p_C) \times (R_A - C_{AC} - C_{BD}) >$$
$$p_C \times (R_A + V_{BC} - C_{AD}) + (1 - p_C) \times (R_A - C_{AD})$$

⇔
$$p_C \times (V_S - C_{AC} + C_{AD}) + (1 - p_C) \times (C_{AD} - C_{AC} - C_{BD}) > 0$$

⇔
$$(C_{AD} - C_{AC} - C_{BD}) + p_C \times (V_S + C_{BD}) > 0$$

⇔
$$p_C > \frac{C_{AC} + C_{BD} - C_{AD}}{V_S + C_{BD}} \tag{5}$$

Equation (5) states that for the foreign firm, cooperation is more profitable than defection if the probability that the local firm cooperates exceeds a critical value that follows from the general structure of payoffs. This critical value is computed by $[(C_{AC} + C_{BD} - C_{AD}) / (V_S + C_{BD})]$. This result is comparable to the result from Section III-1-2-2. However, a strategy does not have to be strictly dominant in order to be chosen. [41]

One can therefore argue that *given* a certain pay-off structure, the higher the probability that the local firm cooperates, the higher will be the readiness for knowledge sharing by the foreign firm. However, a high cooperation probability by firm B does not automatically trigger cooperation by firm A if the general structure of payoffs does not favour cooperation. Quite the contrary: if the dilemma payoff structure presented in Section III-1-1 holds (UD > MC > MD > UC), $p_C = 1$ will allow the foreign firm to defect and to profit from the high payoff for unilateral defection UD.

---

[41] It is straightforward to show that when no dominant strategy exists (i.e.: $C_{AC} - V_S <= C_{AD} <= C_{AC} + C_{BD}$, then the probability p_C is between 0 and 1. For example, higher expected synergies allow for lower cooperation probabilities by the partner to make cooperation the better strategy on average. With the same kind of reasoning, higher costs of giving up technology require higher cooperation probabilities, while high image costs of defection lower the critical probability needed to legitimate cooperation.

The Relevance of Trust

The decision analysis above can be related to literature on *inter-organisational trust*. We therefore take an excursus to briefly elaborate on relevant findings before expressing the mechanisms discussed so far as empirically testable hypotheses. Trust is defined by Curall and Inkpen (2002, p.486) as "a manifestation of reliance, i.e. the decision to take action that allows the trustor's (the actor trusting) fate to be determined by the trustee (the actor trusted)".

The impact of trust on a firm's actions is subject to discussion. There is agreement on the fact that mutual interdependence in cooperative arrangements creates exposure to a partner's potential opportunism (Freeman 1987, Reich and Mankin 1986). However, there exist opposing views regarding the actual behaviour in inter-firm relationships. On the one hand, the 'contractual man' is assumed to be boundedly rational and to act opportunistically, reflecting 'human nature as we know it' (Williamson 1985, Knight 1965, p.270). A foreign firm would thus need to expect the local firm to defect whenever this represents its most profitable strategy and would do the same.

On the other hand, scholars such as Barney (1990), Bromiley and Cummings (1991), Granovetter (1985), and Hill (1990) argue that alliance partners tend to rely on the record of its counterpart's cumulative past behaviours as a guide to future behaviour and to use reputation as a proxy for knowledge of opportunistic intentions. Specifically, when actors have a history of direct interaction, *trust* may emerge on a multitude of levels, whether interpersonal or on an organisational level (Ring and Van de Ven 1992, Curall and Inkpen 2002).[42]

Taking these considerations into account, we argue that a firm has a higher likelihood of cooperating with another firm if the mutual level of trust is high.

---

[42] As a related concept, Kale et al. (2002) develop the notion of *relational capital*, which refers to the level of mutual trust, respect, and friendship that arises out of close interaction at the individual level between alliance partners. Kale et al. (2002) show that building relational capital can help firms to simultaneously achieve alliance objectives, mainly due to the enhanced level of cooperation. Also, relational capital fosters inter-organizational learning, especially the acquisition of difficult-to-codify competencies previously only available to the alliance partner.

Referring to the setting of German-Chinese joint ventures in China, a German investor firm will endow a joint venture with more sophisticated technological capabilities the higher the level of mutual trust within the JV relationship.[43]

The arguments discussed so far can be translated into testable hypotheses when applied to the case of technology transfer within international joint ventures, specifically the transfer of technology by German firms to joint ventures in China. Hypotheses 1 a-d relate to the individual components of the cost structure that have been discussed as relevant for the foreign firm's decision to cooperate or defect, namely C_{AC}, C_{AD}, C_{BD} and V_S. Hypothesis 2 relates to the foreign partner's assessment of mutual trust within the international joint venture.

Hypothesis 1a: A JV's technological endowment by the German investor firm is negatively related to the costs of its own cooperation (C_{AC}).

Hypothesis 1b: A JV's technological endowment by the German investor firm is positively related to the costs of its own defection (C_{AD}).

Hypothesis 1c: A JV's technological endowment by the German investor firm is negatively related to the costs of being unilaterally defected upon (C_{BD}).

Hypothesis 1d: A JV's technological endowment by the German investor firm is positively related to the benefits of mutual cooperation (V_S).

Relating to mutual trust, the following hypothesis is to be tested:

Hypothesis 2: A JV's technological endowment by a German investor firm is positively related to the level of mutual trust between the JV partners.

The attentive reader will notice that although the previously discussed theoretical considerations analyse the decision on whether to exchange knowledge or not as a yes/no decision, the hypotheses relate to the *degree* of a firm's investment in local technological sophistication to exchange. Bruck

---

[43]   Trust between the partner firms has been identified as a success factor for Joint Ventures in China by Mohr and Puck (2003).

(1996) has elaborated on a similar issue and argued that a degree of exchange that gradually is realized basically consists of many yes/no decisions that could be analysed individually. We also argue that the insights provided by our theoretical arguments are only used as heuristics for the derivation of hypothesis, not for the prediction of exact effects.

The next section will extend the theoretical assessment by taking account of repetitive interaction among firms - this allows for the mechanism of *reciprocity,* which has been shown to be responsible for the emergence of cooperation. It also presents arguments for gradually increasing degrees of technological investment over time.

1.2.4. Taking Repeated Interaction and Reciprocity into Account

Up to now, we have described the technology transfer decision of a German technology firm as a 'one-shot game'. The assumption was that once the decision to cooperate is taken, a firm will transfer all of its unique knowledge to its partner. We have also assumed that both firms take this decision simultaneously and that one firm cannot observe the action of the other firm at the time it takes his own decision. However, in cooperations among firms that involve the transfer of technology or other knowledge, the cooperative behaviour can better be characterised as a step-wise interplay between the partners (Bruck 1996). The knowledge transfer is realised incrementally and one player can observe the other player in his behaviour. In the following paragraphs, we rely on basic literature on repeated games (Gibbons 1992a, Gibbons 1992b, Osborne and Rubinstein 1994, Dutta 2000) and then later add insights from other strands of literature in order to derive additional hypotheses for the empirical investigation.

Effectively, technology cooperation can be described as a repeated (or 'iterated') game. Repeated games are characterised by the (in-)definite repetition of a certain game, which in this setting is usually called the *stage* game (Dutta 2000). The strategy of firm A at time t then depends on the observed behaviour of firm B at t_{-1}, t_{-2}, t_{-3}, ... etc., and, more importantly, on the future expected pay-off derived from the interaction with that same firm.

112

With each stage game equal to the game described in Figure 23, the 'new' game that emerges as a result of repetition can have a very different equilibrium outcome than the stage game. A representation of iterated games is shown in the following figure. Each column represents a knowledge-sharing game as described in Figure 23. The foreign firm can not only react to the behaviour of the local firm in previous stage games, but can also take into account its future actions. In the case below, the foreign firm's strategy is to choose cooperation (C) in the current stage whenever the local firm cooperated in the previous stage and defect (D) if the local firm has defected in the previous stage.

Figure 25: Repeated Games

	Stage 1	Stage 2	Stage 3	Stage 4	Stage 5	Stage 6	Stage 7	....
Local Firm	C	C	D	D	C	C	C	....
Foreign Firm		C	C	D	D	C	C	C

Source: Margin et al. (2003), p. 130.

The payoff of the foreign firm (Π_{Total}) is then computed as the sum of the individual stage game pay-offs over all periods, usually taking into account the timing of pay-offs by using a discount factor F (Dutta 2000). That is:

$$\Pi_{Total} = \Pi_{Stage1} \times F + \Pi_{Stage2} \times F^2 + \Pi_{Stage3} \times F^3 + ... \tag{6}$$

The behavioural implications of the so-called 'shadow of the future' are manifold (see e.g. Heide and Miner 1992, Parkhe 1993). According to Dutta (2000), if a player believes that "no good deed today will go unrewarded tomorrow" and, on the other hand, "no bad deed today will go unpunished tomorrow", future pay-offs enter the profit function of players. The effect of reciprocity can cause the emergence of cooperative behaviour, even if non-cooperation is the dominant strategy in single-play situations (Rapoport and Chammah 1965, Axelrod 1984, Oye 1986).[44]

---

[44] How the iteration of games can alter the optimal behaviour towards other players has prominently been demonstrated by Axelrod (1984). In a round-robin computer tournament playing the prisoner's dilemma, the winning computer program was one that embodied the strategy 'TIT FOR TAT', which consists of "starting with cooperation, and thereafter doing what the other player did on the previous move". As

113

The 'cooperation-enhancing effect' of reciprocity applies to strategic alliances, because each firm compares the immediate gain from cheating with the possible sacrifice of future gains that may result from violating an agreement (Tesler 1980).

Application of Technology Transfer to IJVs and Hypotheses

Building on this brief description, we will now introduce the relevant structural characteristics that can be used to predict knowledge sharing behaviour, and specifically can lead to the derivation of hypotheses within the setting of our study. Our arguments relate to the literature on repeated games as well as related strands of literature.

The first structural characteristic refers to the *time horizon of the interaction*. Logic can be used to show that in repeated games where the individual stage game leads to defection (as is the case in the Prisoner's dilemma), rational players will end up defecting in every stage game if the end of interaction is known. Players use backward induction to infer that there will be mutual defection in the last stage game, and thus defection in the second-last stage game, and so forth (Dutta 2000, Feess 2004).[45] For cooperation to occur, a certain probability must exist that the interaction continues indefinitely. As Tesler (1980, p.44) argues, "self-enforcing agreements are not feasible if the sequence of occasions for transactions has a definite known last element. Although termination is certain to occur sooner or later, when this happens must be uncertain in order to sustain a self-enforcing agreement."

Applying this insight to firm behaviour, Parkhe (1993) argues that most inter-firm alliances parallel iterations of unknown length, since they are created with

---

Heide and Miner (1992) describe, the TIT FOR TAT strategy works effectively because it permits players to reward and punish each other. As this example illustrates, cooperation can emerge as a strategy where defection is the equilibrium outcome of the stage game. In fact, it has been shown that a player with a grim trigger strategy ('I start by cooperating, but will defect forever if you defect') can enforce almost all behavioural cycles on another player whenever the time discount factor is sufficiently high (Folk Theorem, see Friedman 1971).

[45] However, the reliability of backward induction for the prediction of players' behaviour has been challenged in cases where the rational solution requires many iterations of backward induction (Basu 2007).

no terminal date.[46] Thus, he concludes that the level of cooperation between partners may be expected to be positively related to the length of the partners' time horizons. This in turn is a positive function of the intended duration of the cooperative relationship and the perceived likelihood of the alliance lasting through the intended duration.

International joint ventures have been described as rather unstable (Inkpen and Beamish 1997, Kogut 1988). Empirical studies have found instability rates as high as 50% (Bleeke and Ernst 1991, Kogut 1988a): joint venture contracts are either terminated before the contractual end date or replaced by new contracts with the same partner, which make the duration of a JV partnership effectively uncertain.

We thus argue that the age of any given IJV can be used as a proxy for past cooperative behaviour, as older JVs have obviously outlived relatively more, failed, JVs within their 'birth cohort' than their 'young' counterparts. On the other hand, the total length of a JV contract can serve as an indicator for the length of an intended interaction and therefore the extent to which both partners are ready to commit resources to the common venture. This leads us to the formulation of the following hypotheses:

Hypothesis 3a: A JV's technological endowment by the German investor firm is positively related to the age of the joint venture.

Hypothesis 3b: A JV's technological endowment by the German investor firm is positively related to the total duration of the joint venture contract.

Furthermore, we refer to the characteristic of alliances to be cooperations of unknown length – with a positive probability that the next interaction will be the last one. In this case, it has been shown that the 'weight of future payoffs' is relevant for the emergence of cooperation (Axelrod 1984, Parkhe 1993, Dutta,

---

[46]   One could also argue that the cooperation periods span such long periods and so many interactions that backward induction is not used by the players (see last footnote).

2000). This weight is represented by the discount rate[47] as well as the time horizon for investments in general. The higher this weight, i.e. the lower the discount rate and the longer the time horizon used to evaluate investments, the more it is likely that a firm will engage in cooperative behaviour.[48]

The same should hold for German firms who invest in JVs in China. We therefore suspect that JVs that are evaluated with low discount rates and long time horizons should exhibit more commitment by the German firm regarding its investment in sophisticated technology than those being evaluated by a high discount rate. The two derived hypotheses are:

Hypothesis 4a: A JV's technological endowment by the German investor firm is negatively related to the discount rate that is used to evaluate the joint venture.

Hypothesis 4b: A JV's technological endowment by the German investor firm is positively related to the time horizon that is used to evaluate the joint venture.

The next structural elements that affect the shadow of the future are related to the *functioning of reciprocity*: the frequency of interaction and behavioural transparency. One shortcoming of the analysis so far is that it requires deviations from cooperative behaviour to be perfectly observable. Also, it assumes that once a deviation has been registered by the other party, punishment will occur immediately. In many contexts these two assumptions are unrealistic, because other players may not have precise information on what a partner has done in the past (Dutta 2000). The relaxation of these two assumptions introduces two more important structural variables, i.e. the frequency of interaction and behavioural transparency (Axelrod and Keohane 1986).

---

[47]   Companies use discount factors to evaluate long-term investments, and these factors vary according to factors such as a project's risk and a firm's cost of capital. The discount rate applied is usually the weighted average cost of capital (r_{WACC}) and effectively assigns a relative weight to future cash flows of an investment project (Ross et al. 1999).

[48]   This is because 'defection' offers quick wins compared to 'mutual cooperation'. However, defection triggers defection by the other partner in future interactions, so that the initially defecting player trades a 'quick win' for the future loss that occurs when mutual cooperation turns into mutual defection.

Frequent interaction between players has the consequence that little time elapses between mutual assessments of the each other's behaviour. A 'defector' can thus not get away very long before he is detected (Axelrod 1984). Since both sides know that an exploitative move can be met with a reciprocal defection in the next stage, the incentive to cheat is reduced (Schelling 1960). *Behavioural transparency* has a similar effect. Effective recognition and control capabilities – the ability to distinguish between cooperation and defection by others and to respond in kind – form the bedrock of reciprocity strategies like TIT FOR TAT (Parkhe 1993). Imperfect information, which may result from (1) observation lags, reaction lags, or both, and (2) an interaction of randomness and inability to monitor the firm's choices (Spence 1978), can impede that effectiveness. Parhke (1993, p. 801) summarises that "[..] frequent interactions and high behavioural transparency encourage reciprocal behaviour. Thus, severally and jointly, these factors lengthen the shadow of the future and promote cooperative outcomes."

A related argument is that frequent contacts between actors favour the sharing of information and the emergence of a set of behavioural norms, informal rules that make it possible to supervise and regulate behaviour (Granovetter 1985). We summarise our considerations with the following two hypotheses:

Hypothesis 5a: A JV's technological endowment by the German investor firm is positively related to the frequency of interaction among the partners.

Hypothesis 5b: A JV's technological endowment by the German investor firm is positively related to the behavioural transparency of the relationship.

Our final two hypotheses relate to the *structural embeddedness* of the JV contract in question (Granovetter 1985), especially the existence of *multiple links* and the *expectation of future transactions* aside from the JV contract under observation. As Curall and Inkpen (2002) argue, globally-acting firms often maintain multiple links to each other. Gulati (1995, p.644) argues that the emerging "social network of indirect ties is an effective referral mechanism for bringing firms together". Via network relations, actors can obtain the information on whether other actors have behaved well in the past or not. Multiple or future links can enforce cooperative behaviour because a 'revenge' to unilateral defection is considerably more severe when a partner has the opportunity for 'multiple answers' (Gulati 1995).

For the case of German JVs in China, this factor has been considered very relevant by interviewed experts.[49] On the one hand, a German firm might have several common JV projects with the same Chinese partner.[50] On the other hand, a German investor might deal directly with the Chinese government regarding multiple projects. In either case, a defection in one partnership might entail a negative response in multiple common projects.

We therefore argue that the more emphasis is placed on links to the current interaction partner aside or after the JV contract under observation, the more pressure for cooperation is created:

Hypothesis 6a: A JV's technological endowment by the German investor firm is positively related to the existence of multiple links to the JV partner (aside from the JV in question).

Hypothesis 6b: A JV's technological endowment by the German investor firm is positively related to mutual dependency after the JV contract ends.

Although both insights from game-theoretic research as well as other strands of literature on knowledge exchange in cooperations offer more starting points for the derivation of hypotheses,[51] we focus our empirical research on the validation of the hypotheses elaborated so far. The following table reproduces all hypotheses again:

---

[49]   Expert interview 5.

[50]   DAX firms such as Siemens AG and Henkel AG maintain over 10 JVs in China, often several JVs with the same partner (see AHK Database for German firms in China referred to in Chapter IV).

[51]   For example, Parkhe (1993) argues for the effect of the total number of parties in a Joint Venture agreement. However, in our study, the vast majority of Joint Ventures only include two firms, which makes the empirical validation of Parkhe's argument difficult.

Table 11: List of Hypotheses

No.	Description
1a	A JV's technological endowment by the German investor firm is negatively related to the costs of its own cooperation (C_{AC}).
1b	A JV's technological endowment by the German investor firm is positively related to the costs of its own defection (C_{AD}).
1c	A JV's technological endowment by the German investor firm is negatively related to the costs of being unilaterally defected upon (C_{BD}).
1d	A JV's technological endowment by the German investor firm is positively related to the benefits of mutual cooperation (V_S).
2	A JV's technological endowment by a German investor firm is positively related to the level of mutual trust between the JV partners.
3a	A JV's technological endowment by the German investor firm is positively related to the age of the Joint Venture.
3b	A JV's technological endowment by the German investor firm is positively related to the total duration of the joint venture contract.
4a	A JV's technological endowment by the German investor firm is negatively related to the discount rate that is used to evaluate the joint venture.
4b	A JV's technological endowment by the German investor firm is positively related to the time horizon that is used to evaluate the joint venture.
5a	A JV's technological endowment by the German investor firm is positively related to the frequency of interaction among the partners.
5b	A JV's technological endowment by the German investor firm is positively related to the behavioural transparency of the relationship.
6a	A JV's technological endowment by the German investor firm is positively related to the existence of multiple links to the JV partner (aside from the JV in question).
6b	A JV's technological endowment by the German investor firm is positively related to mutual dependency after the JV contract ends.

Source: Own computations.

This list of hypotheses is subject to investigation and testing by means of the quantitative study described in the following chapter. Before we proceed to describe the quantitative study, however, we review existing literature for other, general, factors that can be expected to influence the technological endowment and the relative 'technological commitment' of a German JV partner in China.

2. Discussion of General Influence Factors

Given that the degree of technological knowledge that a foreign firm transfers to an IJV may be the result of factors such as technical feasibility, regulatory restrictions or economic forces, the leeway for adjusting this degree as a result of technology-sharing considerations can be quite limited.

As a supplement to the theoretical considerations derived in the previous chapter, we thus dedicate the following paragraphs to an extensive review of previous literature on reasons why international firms invest in technological endowment abroad, and specifically within international joint ventures. Because this question is rather broad, the review includes findings from a variety of fields. Insights are drawn both from a literature review as well as from expert interviews that have been conducted as part of this study.[52] Our resulting review takes account of the most relevant factors influencing the technological endowment of an organisational activity abroad, although we do not claim this review to be comprehensive.[53]

Two main factor categories are used: *external* influence factors and *internal* factors. While *external* factors are considered to be those that are not subject to the influence of the investor firm (but affect it nonetheless), we use the term *internal* for such factors that are *firm, product, project, and technology specific* and are possibly subject to the foreign investor's influence.[54]

---

[52] See Appendix 3.

[53] A firm's behaviour with regard to technology transfer depends very much on the nature of a firm's activities and the motives of the activities in a foreign market. Unless stated otherwise, we focus our attention on cases where the primary motive is to serve a foreign market with locally manufactured and/or adapted products.

[54] We draw on prior frameworks for frameworks regarding this question. Baranson (1970, p. 435) argues that the feasibility of technology transfer is influenced by four groups of factors, namely (1) the complexity of the product (2) the transfer environment of the donor and the recipient country (3) the absorptive capability of the recipient and (4) the transfer capability and the profit-maximising strategy of the firm. Prior research in the field of internationalisation of R&D has investigated the relevance of a firm's external environment, distinguishing between supply-side drivers and demand-side drivers (Granstand et al. 1993, Dunning and Narula 1995, OECD 1998, Gassmann and Zhen 2004). Beckmann and Fischer (1994) develop a refined scheme that includes four categories of factors (input-oriented, output-oriented, efficiency-oriented, and political/social-cultural). Finally, Gassmann and Zhen (2004) use a classification scheme comprising three factors, namely input-oriented motivations, performance-oriented motivations, and business-ecological motivations when discussing the factors that drive firms to invest in R&D capability in China.

2.1. External Influence Factors

'External' influence factors are factors that affect the foreign investor firm and its behaviour regarding technology transfer without being subject to the foreign firm's influence. On the one hand, we consider factors that pull technology into the foreign market by exerting pressure on the company's output or the way it is able to sell it. Products or services have to comply with local requirements, customer requirements and regulations. This compliance may induce a firm to transfer technological capabilities into the foreign market (Beckmann and Fischer 1994). On the other hand, we consider factors that are connected to cost advantages that the target market might offer.

2.1.1. Local Market Conditions

When local market requirements significantly differ from home country requirements, companies need to deploy local development capability to that target market (unless they serve the market from another location). This is because technological knowledge overseas helps companies obtain local knowledge and locally adapt their products (Howells 1990, Kogut and Chang 1991, Hakanson and Nobel 1993, Beckmann and Fischer 1994). Of course, the relevance of these factors generally depends on the extent to which local market conditions require an adaptation of a firm's output and the degree of economic pressure exerted.

The origin of country-specific requirements is generally regarded as consisting of a mixture of socio-cultural, economic, and natural differences between countries. Socio-cultural differences exist when customers have different value systems or patterns of use for certain products. Economic reasons for differences include the average purchasing power of the population and the distribution of income. A foreign country can also have different climatic conditions, and/or geographic characteristics. Even the population living in that country may have specific requirements for a company's output (Beckmann and Fischer 1994).[55]

---

[55] For example, the use of shampoo in a given country can be influenced by the hair-washing habits of the population or the belief of what determines a person's beauty,

In the case of China, it is obvious that customer requirements, value systems, economic conditions, and even natural factors deviate widely from German or European conditions. Applying his well-known cultural dimension scheme, Hofstede (1980, 1994, 2003) consistently finds large cultural differences between Germany and China, as the following table shows.

Table 12: Scores of Germany and China in Hofstede's Cultural Dimensions[56]

Country	PDI	IDV	MAS	UAI	LTO
Germany	35	67	66	65	31
China	80	20	66	30	118
World Avg.	55	43	50	64	45

Source: Hofstede (1980, 2003).

One should thus expect firms to respond with investment in local knowledge and technological capabilities in order to adapt their output. Gassmann and Zhen (2004) argue that in the case of China, one main reason why companies establish development bases there is to develop products locally specifically for the Chinese market. In a similar fashion, research results of Li and Zhong (2003) show that between 1995 and 2000, the major motive of foreign firms to enter R&D alliances in China (including equity-based and non-equity based alliances) was local adaptation.

With respect to the research question at hand, we can expect differences in local market conditions to be one major reason for firms to equip activities in China with technological knowledge. Although the study looks at activities of German firms in China only – implying that the country specific market requirements and cultural differences don't vary across cases – there are specific factors that can *affect* firms to a varying degree. These are customer-specific requirements, compliance with output regulations, and the susceptibility to IP theft and the

---

the population's income distribution and the hair properties of a population. In the EC merger case involving the acquisition of Wella GmbH by Procter & Gamble in 2003, the European Commission found all European countries to represent different national markets for the supply of hair care products. Source: European Commission (2003).

[56] Abbreviations: PDI: Power Distance Index, IDV: Individualism, MAS: Masculinity, UAI: Uncertainty Avoidance Index, LTO: Long-Term Orientation.

related sensitivity regarding a country's IP regime. These factors will be discussed in detail below.

Customer-Specific Requirements

As Beckmann and Fischer (1994) put it, the most important transaction partner that affects the internationalisation of R&D for a firm is the customers. Products must in many cases not only be adapted to the market in general but also must meet the requirements of specific customers (Root 1994). Typical industries that require customer-specific R&D efforts are the chemical industry, the automotive component sector, or power generation (Beckmann and Fischer 1994, expert interview 4). Second, firms invest in local R&D to allow for a timely adaptation of products that require customer and market-specific accommodations (Gassmann and Zhen 2004). As noted by an expert on IJVs in China, 'speed' is a crucial element for serving Chinese customers and that is why firms need local technological capabilities.[57] Finally, for many B2B supplier-client relationships, the importance of after-sales service such as IT maintenance, process engineering, or technical maintenance is very high (Beckmann and Fischer 1994). Offering these services requires technical capabilities and qualified personnel close to the customer.

Summarising the arguments above, customer-specific requirements should be expected to be among the most important factors driving technology into a foreign market.

Compliance of Output with Regulations

A firm's output – whether product or service – must comply with a country's regulations. These regulations can force firms to establish development capacity in China (Gassmann and Zhen 2004). Because product regulations tend to be highly industry-specific they can affect certain investors to China while not affecting others.

Several industries can be mentioned as examples. As Beckmann and Fischer (1994) note, the pharmaceutical industry, cosmetics and phytosanitary products

---

[57]   Expert interview 5.

often need to adapt a product to country-specific regulations for gaining approval by local authorities. Another industry that is quoted as subject to local regulation is the automobile industry, because almost all national authorities maintain different regulations for the approval of automobiles (Kasperk et al. 2006). Another example is the food industry: national food administration authorities maintain different regulations on which food ingredients are allowed. This requires international food manufacturers and their suppliers to develop country-specific solutions.[58]

Finally, industries in China that are sensitive because of *national security* are usually governed strictly (expert interview 4 and Section II-4-2). Not only are strict safety standards enforced, but also the transfer of technology is specified in detail.

On the whole, one should thus expect that firms who supply a product that is subject to tight regulations should have a high propensity to invest in local technological capabilities.

IP Protection Regime and Susceptibility of a Firm's Technology to IP Loss

As Gassmann and Zhen (2004) describe, a factor that generally limits an investor's willingness to deploy technological knowledge to China is a perceived lack of IP protection. This lack of IP protection stops foreign firms from importing their core technologies, research, or equipment to China. Mansfield (1995, p. 26) records the quote of a general manager of international operations of a major Japanese electrical equipment producer stating the following:

> *"If the country does not have an adequate system or practice to protect intellectual property rights, our technology transfer to that country will be limited and we protect our rights in the following way. We will ensure, in the technology transfer agreement, that the important or core technology will not be transferred without adequate consideration. For example, necessary parts will be supplied from our factory"*

---

[58] For example, the yellow food colour raw material 'saflor' is allowed in the EU, while being prohibited in the U.S. The ingredient 'monascus' is allowed in Japan but not in the EU. See www.natcol.org.

For China, a major reason of foreign investors' reluctance is uncertainty regarding intellectual property (IP) conflicts. As it is perceived, starting and winning a patent case in China is still almost impossible and definitely time consuming. Even if the foreign partner has clear evidence of IP theft, their means of counteracting are limited due to what is seen to be strong protectionism by the government (Gassmann and Zhen 2004).

However, China's entry into the WTO and other developments (described in Section II-4-2) are incrementally changing this perception. As a result, the number of foreign applications for patents attributable to high tech industries is continually rising (Gassmann and Zhen 2004).

Although a national IP regime itself is constant for the whole of China, one can take account of the German investor firm's *susceptibility* of to IP loss and the resulting perceived risk of IP theft incurred by the German firm.

2.1.2. Industry-Specific Market Attractiveness

Next to differing local market conditions, a factor that is generally held responsible for technological investment into a foreign country is the attractiveness of a market. As described by Fu (2005), the two factors that are generally regarded to make a market attractive are the *size* of the market and the expected *growth rate* of that market.

A high expected sales volume in a given market correlates with high market-specific investments, because investments in foreign operations usually represent sunk costs in tangible or intangible assets (Terpstra and Yu 1998). As a consequence, high expected sales promise faster amortisation of such an investment as well as economies of scale during operations (Fu 2005). Investments in technological capability are usually sunk costs that do not vary significantly with sales volume and are thus encouraged by a high market attractiveness (Expert interview 3).

China is widely regarded as a large and promising market. As Gassmann and Han (2004) mention, the recent growth of the Chinese national economy and its overwhelming market size has positioned China amongst the most important markets for TNCs. Because there might exist differences among individual industries within China, our study tests for the *industry-specific market attractiveness* on the propensity of German firms to invest by transferring technology to the market. We expect that the higher the firms perceive the attractiveness of a targeted market segment in China, the higher is the propensity of firms to invest in local technological capabilities.

2.1.3. Differences in Cost Levels

The next factor we discuss is the difference in cost levels and how they can influence a firm's propensity to transfer technology abroad. We introduce two arguments that point in different directions.

On the one hand, comparatively low cost levels in a country might induce an investor to dislocate activities to a foreign country (Vernon and Davidson 1979, Hobday 1995, Bennett et al. 2001, ITEM/TECTEM 2005) in order to realise cost-savings. Gaining access to low labour and raw material costs is a common motive for firms to invest in China (Roehrig 1994). In a study of 20 EU industrial firms reporting on 57 recent technology transfer projects to China, the realisation of a cost advantage had been stated in 57% of the cases (Bennett et al. 2001).

However, the implications of the cost-cutting motive for the sophistication of an IJVs technology endowment are ambiguous. If 'cost-cutting' is a foreign site's only purpose, then the technological level of operations should be rather low, as studies regarding off-shoring show (Deutsche Bank 2005). On the other hand, cost-cutting in technologically demanding activities and even R&D have gained importance in the context of international market entry. Because the salaries for qualified technical personnel vary significantly across countries and these salaries represent on average about two-thirds of the costs of R&D activities (Beckmann and Fischer 1994), international firms start to locate their R&D activities in countries were such salaries are low.

The cost advantage to be achieved in China can be considerable. Although the wages of highly qualified Chinese R&D staff are high compared to the average income/capita, they still average between 20% and 25% of the R&D staff salary in triad regions. Running similar R&D facilities in China costs about only one-tenth of what it would cost in the USA (Li and Zhong 2003). According to this argument, firms could locate some technologically challenging activities in China due to low cost levels.

Taking into account both arguments above, it is hard to predict if low local cost levels lead to high technological endowment of an international joint venture: if 'cost-cutting' is a JVs sole purpose, then the technological level of operations should be rather low. If low cost levels encourage the investment in R&D capacity, one would expect high levels of technological sophistication.

2.1.4. Government Incentives

In the case of China, a factor that has a major influence on the technology transfer behaviour by foreign investors is the public policy regarding government incentives. As described in Section II-4-2, the Chinese government follows a strategy of exchanging 'technology for market'. In order to implement this strategy, the government subsidises technologically advanced activities on its territory (Gassmann and Zhen 2004) and even actively gets involved by offering specific incentives.

We expect that the higher the governmental incentives given to the investor, the higher the commitment to invest in local technological capabilities. One particular mechanism in this context is that the Chinese government includes the commitment for technology transfer in public procurement contracts to create an auction-type environment for the foreign investors: the Chinese government literally 'auctions off' potential market access to a number of 'technology bidding' investors.[59] By stimulating the degree of competition among foreign investors, the Chinese government succeeds in generating strong commitments with regards to investments in local technological capabilities.[60]

---

[59] See the examples described in section II.4.2.

[60] As is known from auction theory (Klemperer 1999), the formal setup of a competitive setting among bidders is able to increase the bid, even if all other factors are held constant.

However, Baranson (1970) argues that a government's insistence on local content can repel foreign investors. An extreme form is the case when market entry by means of a joint venture is prescribed by investment laws and this form does not represent the investor's *preferred option* to enter the target market. Next to the positive effect of government incentives in general, one could argue for a negative propensity of German firms to share technology should the respective set-up be the result of 'enforcement' by the government. We therefore control whether a given joint venture is the investor's preferred entry mode and whether it is a result of government enforcement.

2.2. Internal Influence Factors

The previous section gave an overview of factors that have an external influence on a firm's propensity to transfer sophisticated technology to China. Such factors 'pull' technology into a market or make a firm 'push' technology into the market, but all of them can be viewed as part of the firm's environment, and the firm's influence on them is usually low.

We now discuss 'internal' factors that are *firm- and technology-specific* and are possibly subject to the foreign investor's influence. First, we discuss factors that are related to characteristics of the technology 'sender' and 'receiver'. Second, we discuss factors relating to the feasibility and effectiveness of TT and what factors influence them. Third, we review how the recipient's characteristics and the administrative environment for TT in the recipient country affect technology. This also includes some considerations on the relationship between the technology sender and the technology receiver. [61]

---

[61] This categorisation is closely related to the work of Baranson (1970). He states that the relative cost and feasibility of international technology transfer depends on four sets of interrelated factors: (1) the nature, especially the degree of complexity of the product (2) the transfer environment of donor and recipient country (3) the absorptive capability of the recipient and (4) the transfer capability and the profit-maximising strategy of the technology sender.

We observe at this point that some hypotheses derived in Section III-2 belong to the same general set of factors, particularly the characteristics and motives of the foreign investor, the local investor, and the relationship between them. Due to this overlap, we here only discuss influence factors that are not covered by the main hypotheses.

2.2.1. Characteristics of the Technology Sending Firm

The starting point for our analysis of firm- and technology-specific factors is the firm that acts as the 'technology sender' – in this case the German firm. As the joint venture activities in China represent the German firm's expansion into the Chinese market, it is in most cases the German firm who takes the initiative and pro-actively establishes a joint venture with a Chinese partner as part of its international strategy. Therefore, the actual technology transfer crucially depends on the technology sender's motives, resources, and experiences regarding international technology transfer. The insights from previous studies regarding such factors and their predicted impact on the empirical results are summarised in the paragraphs below.

The first characteristic that we consider is a firm's *resource base*. As has been shown by Teece (1977), a technology transfer project to any kind of foreign operation involves a significant commitment of resources.[62] The formation of joint ventures requires especially high efforts, e.g. due to the partner finding process, legal negotiations, and IP protection measures, which is why resources can be a limiting factor. Because the *size* of a firm is a good proxy for its resource base, we expect that the ability to share technological knowledge with a joint venture partner increases with firm size.[63]

---

[62] Important cost factors are: investing in specific machinery, dispatching qualified personnel, supporting the learning process of local employees by means of training courses and coping with initial excess manufacturing costs. Results of a study by Teece (1997) show that a) pre-engineering technological exchanges; b) engineering costs for process design; c) R&D personnel during all phases of a project; and d) start-up training and "excess-manufacturing" can constitute up to 59% of total project costs.

[63] Of course, the fact that only companies with *existing joint ventures* in China are targeted by our survey, the size argument is somewhat weakened, because the firms

A second relevant characteristic of the investor firm is its *degree of internationalisation*. A firm whose assets, sales, and employees are already spread across the world is more likely to be able and willing to localise its technological assets in China than a firm that is mostly dependent on its domestic market. Therefore we expect that the more *transnational* a firm is, the more likely it is to make strong commitments with regard to technology investments in China.[64]

Third, for a firm's ability to transfer technology abroad, its *experience in ITT projects* matters. As noted by Teece (1977), technology transfer is a decreasing-cost activity and the more experience a firm has in this activity and the more often it has engaged in projects transferring a particular technology before, the fewer resources it will need to repeat that effort. A firm with previous experience regarding ITT to China should therefore exhibit confidence regarding that specific target market and comparatively strong commitment regarding technological investments.

Fourth, a firm's *global strategy* matters.[65] Bartlett and Goshal (1998) distinguish two major types of benefits deriving from worldwide operations: benefits from *local responsiveness* and benefits from the *worldwide coordination and integration* of activities. Applying a two-by-two matrix, the four resulting strategy types are the *international* strategy, the *multi-local* strategy, the *global* strategy, and the *transnational* strategy.[66] This standard framework can be used

---

must have had the necessary resources to establish a JV in the first place. However, we suggest that the *level* of technology transferred depends on firm size, even across existing joint ventures.

[64] Transnationality has been measured by the "transnationality index" ('TNI') as the average of a) the share of foreign assets / total assets, b) the share of foreign sales / total sales and c) the share of foreign employees / total employees (World Investment Report 2004).

[65] In many cases, the degree of transnationality is a direct result of a firm's global strategy, which will be discussed in the following section. In other cases, however, such as in the case of Interbrew or Philips, the transnationality of a corporation can derive from the small size of the historical home market (Belgium and respectively The Netherlands).

[66] An earlier framework of this kind was offered in 1969 by Perlmutter (1969), who distinguishes between *ethnocentric*, *polycentric*, *geocentric* and *regiocentric*

to predict a firm's commitment of technological resources to country markets: the more a firm is committed to local responsiveness and the less it is committed to benefits from global integration, the more one could expect it to invest in the technological endowment of a specific target market such as China.[67]

Fifth, we take into account the *role of the joint venture* as a part of the firm's strategy. Here, it is important to take into account which functional activities a given joint venture is supposed to perform (e.g. sales, production, or sourcing), the exact legal form it has (contractual vs. equity), and the ownership share of the foreign investor. We expect that German investors invest more technology in a given joint venture, the higher is their ownership share and the more challenging are the JV's functional activities. Also, equity joint ventures should receive more commitment than contractual joint ventures.

Finally, one should expect that *competitive pressure* on the technology supplier results in high technological commitment towards a joint venture, because joint ventures allow firms to enter and penetrate a market very fast (Root 1994). One piece of evidence relating to this argument is that technology suppliers subject to competitive pressure are more willing to engage in cooperations instead of sole ventures (Baranson 1970, Vernon 1979, Wells and Stobaugh 1984). Another piece of evidence is that managers of foreign investor firms in China cite peer-pressure among TNCs as a driver for investing in R&D capacity in China (Gassmann and Zhen 2004).

Summarising our considerations above, we expect that firms that are comparatively large, transnational, that follow a multi-local or transnational strategy, have a lot of experience in ITT and also are under competitive pressure to expand to China should be more willing to invest in local technological capacity (all else being equal).

---

companies. However, we consider the framework by Bartlett and Goshal (1998) more appropriate in this case.

[67] Similar predictions can be derived from frameworks on a firm's international configuration of R&D (Beckmann and Fischer 1994, Gassmann and Zhen 2004).

2.2.2. Product and Technology Characteristics

The characteristics of a technology have a fundamental influence on its transferability and thus on the propensity of firms to actually transfer it. In this section, we therefore introduce some key characteristics of products or a technology to be transferred to international joint ventures in China that can have an influence on the 'end result'.

Concerning the properties of technology, we also have to take into account the results of our discussion of Section II-2-2: there is a general bias in the type of technology to be found in joint ventures, because the nature of the technology to be transferred is among the factors that influence a firm's decision on which market entry form to use in the first place. However, we argue that there is considerable variance of the level of sophistication of technologies that are transferred *within* the market entry form 'joint venture'.

A first factor that we take into account is the technology's *age*. As Teece (1977) could empirically prove, the costs of transferring a given technology decline with its *age*: state-of-the art technology is cheaper to transfer than leading-edge technology.[68] This is because over time, the technology sender's knowledge about the technology is extensive, the engineering design has become stable, transfer problems have been ironed out, and employees are familiar with the tacit knowledge of operating it successfully.[69]

The same argument can be extended to an industry level. As Teece (1977) argues, the costs of transfer projects also decrease with the *number of firms in an industry utilising the technology* that is to be transferred: the greater the number of firms using the same or similar technology, the greater the likelihood that a technology is generally available and can therefore be transferred to a target location at lower cost.

---

[68]   The age of technology has here been defined as the number of years since the beginning of the first commercial application anywhere in the world and the end date of a given technology transfer project.

[69]   The age of a technology is closely related to a product's stage in the product life cycle (Vernon 1979), but we do not further elaborate on that concept here.

Finally, the *number of previous times* a certain technology has been transferred by a firm is recognised as a factor driving down transfer costs (Teece 1977). The more routine a firm has with operating and transferring a given technology, the easier it is to engage in an additional technology transfer project: there exists a 'learning curve' on the firm level as well as on the technology level (Teece 1977).

The next group of factors relates to a technology's *inherent replicability*. This characteristic of a technology describes the efforts one has to undertake to replicate it for the sake of employing it in a different location or setting. Deep understanding of a technology is often required to accomplish replication (Teece 2000). Lippman and Rumelt (1982) even argue that some knowledge is not transferable at all, because it is so complex that the firm itself cannot formalise it.

One proxy for replicability is the distinction between *product* and *process* technologies. Since process knowledge is typically embodied and is thus tacit in nature, it cannot be transferred by simply transmitting information (Nelson and Winter 1982, Teece 1982). Polanyi (1958, p. 53) argues that tacit knowledge can be passed on only "by example from master to apprentice", usually through person-to-person demonstrations and instructions. Process technology is thus inherently more difficult to transfer than product technology (Yan and Gray 1994, Frederiksen and Sedita 2005). The expectation we form for our study is that firms employing product technologies might be more likely to transfer a high proportion of their knowledge to a JV in China than firms employing process technologies, simply due to better feasibility.

A second approximation for a technology's replicability is its de-composition into the dimensions 'complexity', 'teachability' and 'codifiability', as used by Kogut and Zander (1993) and de Almeida (1995).[70] Applying the arguments of Kogut and Zander (1993) to our setting, we expect that the less complex, the

---

[70]   Codifiability measures the extent to which knowledge can be articulated in documents. Teachability expresses the ease by which know-how can be taught to new workers. Finally, complexity is defined as the number of critical and interacting elements in an activity.

more codifiable and the more teachable a technology is, the higher is the relative degree of it that is mastered by the joint venture and the local joint venture partner.

A counter-argument to the replicability discussion is offered when one includes the risk of unwanted technology diffusion and the strength of the local IP protection regime (Maskus 2003). For technologies that are inherently easy to replicate and that are not subject to a strong IP protection regime, firms frequently have appropriation concerns (Merges and Nelson 1990, Anand and Khanna 1997, Oxley 1997). Especially if the limits of the technology being transacted upon are difficult to specify legally, firms may not be able to capture the economic rent of their innovations (Gulati and Singh 1998, Anand and Khanna 1997). Firms thus have the concern of unwanted or non-transparent technology transfer and shy away from transferring technology at all (Gassmann and Zhen 2004). As a result, one could argue that the easier it is to replicate a certain technology the lower will be a firm's propensity to share it.

One final factor to consider is a technology's *strategic value to the firm.* A firm generally protects the knowledge that is crucial for its competitive advantage. Technology with a high strategic value is therefore preferably transferred to sole ventures, because this market entry form entails a comparatively low risk of unwanted diffusion (Section II-2-2). Although strategically important knowledge is often tacit in nature, its transfer to a joint venture gives the local partner the possibility to appropriate it (Frederiksen and Sedita 2005). Foreign firms are therefore reluctant to transfer strategically important technologies to an IJV.[71] Another related concept is how *modern* a technology is: the more modern a competence is, the less eagerly a firm will share it, because is likely to be unique to the firm.[72]

---

[71] This expectation extends to concepts such as 'core' vs. 'non-core' technologies. Some of the interviewed firms had a strict practice of classifying technologies into "core" and "non-core". The transfer of "core" technologies abroad or to third parties would then require approval of the organisation's highest levels (Expert interview 8 and Section II-1-1).

[72] Determining how unique a valuable firm resource must be in order to have the potential for creating a competitive advantage is a difficult question. However, the value of a resource decreases with the number of firms that possess it. As Barney

A potential caveat to point out here is that one can consider the strategic value of a technology that ends up in an IJV as a proxy for the transfer efforts by the foreign firm, and thus can interpret it as a *result* rather than an influence factor. This causes problems in identifying the direction of causality in our study (see discussion in Chapter V).

Besides the factors mentioned above, there exist other characteristics that may influence the way firms might transfer a technology, e.g. the size of a final product (Steenhuis and De Bruijn 2005) or the capital intensity of a technology (Teece 1977). However, a discussion of all factors mentioned in the literature is beyond the limits of the present work. We now go on to discuss the final set of factors, namely the ones that relate to the recipient of the technology and practical implementation barriers.

2.2.3. The Technology Recipient and Practical Implementation Barriers

A discussion of factors that influence a foreign investor's technological endowment to an IJV cannot but take into account the properties of the technology recipient, its environment, and – last but not least – the relationship between the technology sender and the technology receiver. This also extends to practical implementation barriers.

Because the derivation of hypotheses in Section III-1 already takes into account many properties of the relationship between the technology sender and the technology receiver, we only discuss those concepts not explicitly discussed by the last sections.

Key to technology transfer is effective communication between those who have developed the technology and those who can use the technology (Rouach 2003). A practical and therefore crucial factor that determines the technological endowment of an IJV is the *absorptive capacity* of the recipient organisation.

(1991, p. 212) puts it: "As long as the number of firms that possess a particular valuable resource (or a bundle of valuable resources) is less than the number of firms needed to generate perfect competition dynamics in an industry, that resource has the potential of generating competitive advantage".

This is the idea that a firm or country needs to have a certain skill level in order to be able to successfully adopt foreign technology (Cohen and Levinthal 1990, Keller 1996). Absorptive capacity has been related to factors such as human capital (Eaton and Kortum 1996, Nelson and Phelps 1966). It can also be related to the extent of R&D spending, because R&D investments are necessary for a firm to acquire outside technology (Cohen and Levinthal 1990). Other factors have been suggested to be the recipient's firm size, its years of experience, and the existing base of manufacturing knowledge (Teece 1977).

Practical implementation barriers to technology transfer between both firms have to be considered. First, efficient communication can be hindered due to a cultural gap and a language barrier.[73] As outlined in Section III-2-1, the Chinese culture and the German culture are very different. Contrary to the German culture, the Chinese culture is a very high-context culture (Hofstede 1994). The resulting culture gap and language barrier can hinder communication significantly, especially hindering daily interactions due to issues like communication style, or "face saving" (Gassmann and Zhen 2004, p. 431). Technology transfer can also be hindered by administrative barriers, e.g. when importing materials and people are difficult. Gassmann and Zhen (2004) note that in the case of China, foreign investors need to invest significant effort into building trust and good working relationships with the Chinese government before local operations of foreign firms can be started. High practical barriers resulting from the recipient firm's host country are therefore expected to reduce the technological endowment of an international joint venture.

Summarising the considerations in this section, we have discussed a variety of external as well as internal factors that could be linked to the (result) of the technology transfer effort by a German joint venture partner in order to endow the local organisation with sophisticated technological resources and capabilities.

---

[73]   As noted by Arrow as early as 1969, the costs of communication are a fundamental factor influencing the worldwide diffusion of technology. However, he mainly refers to the costs of information transfer itself, which are not assumed to be very high anymore.

In combination with the previous section that served to discuss our main hypotheses, we have now offered a comprehensive picture of the alleged mechanisms that correspond with a high commitment by the German technology sender and we are ready to test for individual hypotheses empirically.

The quantitative study described in the following chapter serves to empirically test the hypotheses, controlling for the general factors discussed so far.

Chapter IV: Empirical Study

This chapter is dedicated to the description of our empirical study carried out among German Chinese International Joint Ventures between November 2006 and September 2007. It was designed to allow for the systematic testing of the theoretical hypotheses derived in Section III-1, taking account of the general factors discussed in Section III-2.

In general terms, this thesis alleges that a foreign investor's commitment to transfer sophisticated technologies to an international joint venture is influenced by – among other factors – the co-operative relationship between the foreign investor and its local joint venture partner. This allegation and several derived hypotheses are to be tested by means of our empirical study.

This chapter is structured as follows. Section IV-1 provides an overview of the research design of the study, explaining the choice of the empirical setting, the structure of the investigation, and alternative measurement concepts for the dependent variable. We adopt several alternative scales to measure slightly different concepts that relate to the technological commitment of a German investor to its international joint venture and its Chinese joint venture partner. Section IV-2 then proceeds by presenting the results of the study. It starts out by describing the sample and then presents both descriptive results as well as results from multiple regression analysis for the empirical evaluation of hypotheses. Finally, Section IV-3 summarises and discusses the research results.

1. Research Design

In order to produce meaningful empirical evidence that can be used to evaluate the theoretical framework proposed in Chapter III, this study was designed to measure the degree of technology-sharing commitment that a foreign investor to international joint ventures shows towards the joint venture itself as well as the local partner to the joint venture in question.

The following section describes why the specific setting of German industrial firms in China was chosen in order to accomplish this task. Because the sharing

© Springer Fachmedien Wiesbaden GmbH, part of Springer Nature 2008
M. Hoeck, *Cooperation and Technological Endowment in International Joint Ventures: German Firms in China*, Edition KWV, https://doi.org/10.1007/978-3-658-24355-5_4

of technology is a rather abstract concept, Section IV-1-2 then discusses alternative measurement concepts for the dependent variable in detail.

1.1. Choice of the Empirical Setting and Structure of the Investigation

The empirical setting of choice to evaluate the theoretical questions as well as the hypotheses derived in previous sections is international joint ventures of German industrial firms in China. As the discussion in Chapter II shows, technological sophistication represents an important topic for international firms in both Germany and China. This particularly applies to the industrial sector. The industrial sector is the largest and fastest growing part of the Chinese economy, representing 42% of 1996 gross output. It also received 96% of all foreign technology transferred from 1979 to 1995 by value (Dougherty 1997). The industrial sector and in particular the trade in technologically advanced goods is also a key element for the German economy (BMBF 2006). Consequently, Germany is one of the major (technology) investors in China, as Section II-4-3 shows.

Besides the economic relevance of the topic, it presents a particularly good setting for the investigation of the hypotheses of this thesis. On the one hand, the targeted sample represents a large set of cooperative relationships that are homogeneous with regard to:
- the market entry form international joint venture,
- the geographic target market China, and
- German industrial firms as foreign investors.

On the other hand, the cooperation behaviour with regard to the exchange of technology is expected to be rather heterogeneous. A survey by Fu (2005) among 52 medium-sized German companies in China reports that 40% of joint ventures in the sample are a consequence of governmental regulation. Only 60% of joint ventures in the sample were founded out of managerial considerations alone and can thus be regarded as 'free-choice' co-operations.

Table 13: Influence of Investment Regulations on Market Entry Form in China

MEF Result of Investment Regulations?	Representative Office	Joint Venture	WFOE	Total
Yes	42,9%	40,0%	11,8%	29,3%
No	57,1%	60,0%	88,2%	70,7%
Total	100%	100%	100%	100%

Source: Fu (2005), p. 176.

Given the heterogeneity of this sample, one can expect the delicate issue of technology transfer to be handled quite differently from case to case, a reason for why this dataset seems to be very suitable for the investigation of the hypotheses derived in Chapter III.

Structure of Investigation

For the chosen sample of international joint ventures in China, the co-operative setting as well as the sophistication of a German investor's commitment to invest and share sophisticated technological assets and capabilities is to be measured. The chosen empirical method is multivariate regression analysis based on data from a survey among German firms with current participations in international joint ventures in China.

A preparatory step for this has been a *qualitative study* - structured interviews with a variety of experts on the topic. The following paragraphs briefly describe the structure of both the qualitative study and the survey.

The qualitative study was meant to gather first insights on the topic of technology transfer. This was done by contacting organisations and individuals who are known to have expert knowledge on the topic.[74] During this first assessment of the topic, some relevant issues relating to technology transfer within German-Chinese joint ventures could be analyzed. One focus of this prior study was the question of if and how technology transfer is negotiated between the joint venture partners. Another important aspect to be clarified was the

---

[74] A list of experts and the description of their backgrounds can be found in Appendix 3.

question of how the degree of technology transfer can be measured, given that a relative knowledge transfer assessment was agreed to be possible.

Finally, experts were directly asked for their expectation of the most relevant factors influencing the degree of technology transfer from German industrial firms to their joint ventures with local partners in China. Results of the qualitative study have already been mentioned throughout Chapters II to III wherever appropriate. However, the main objective for the expert interview was the preparation of the large scale survey among German JVs in China.

The second step involved a *survey* of senior managers responsible for German-Chinese joint ventures in China. Target firms were all German industrial firms who were currently invested in a joint venture in China for the local production and/or sale of goods.[75]

The potential survey respondents were managers within the target firms with expertise on the history and the current status of the firm's joint venture(s) in China. As this knowledge is not strictly attributable to a single function within a company, the respondents had a variety of different backgrounds and were generally high-ranking senior managers.

The following information sources were used to identify and contact potential respondents:

- The database of German firms in China provided by the German Chamber of Commerce in China (AHK Beijing), comprising a total of over 2900 registries.. Next to the name and industry type, the list gives the contact information of the German member of the Chamber of Commerce. Also, background information on the joint venture itself, such as date of founding, number of employees, and invested capital are provided.
- The 'Firm Information System' (FIS) of the German Chamber of Commerce. This database provides information on German firms and

---

[75] A single joint venture can host several lines of products, which might be related to different technological knowledge. Therefore, we asked on the level of the individual product. In cases of several products per joint venture, we gave the respondent the freedom to choose one or more selected products to report on.

their foreign activities and was used to complement and verify the information provided by AHK Beijing.

- Direct contact of managers through visits to management congresses such as the EuroMOT 2006 and the R&D Management Conference 2006.

The combined information sources resulted in a list of 259 international joint ventures by German firm whose existence could be confirmed (or at least not falsified). This represents only a portion of the 484 equity joint ventures registered by the databases quoted above (and Section II-4-3), because a number of firms could not be reached, some database entries could not be confirmed or existing joint ventures had been sold or turned into other market entry forms (mostly WFOEs). The exact size of the 'population' of German-Chinese joint ventures remains unknown to the author.

Using this empirical setting of German-Chinese joint ventures as a background, the quantitative study was designed to enable the extraction of strategic considerations regarding knowledge transfer and sharing by technology suppliers. A customised survey needed to be conducted because the sophistication of technology transfer is very difficult to measure quantitatively from generally available data (Dougherty 1997). The survey was carried out between October 2006 and September 2007.[76]

For several reasons, this survey had to rely on the subjective assessments by respondents of the German investor firms (i.e. only the foreign investor's perspective). Although an ideal study design would have considered both the German as well as the local Chinese partner as a source of information, this was considered to be nearly impossible to realize for the topic of technology transfer. Potential respondents of the Chinese partner were not identifiable provided the resources at hand. Furthermore, it was assumed that an objective and informative response to a questionnaire was extremely unlikely to be obtained - given the sensitivity of the topic. Potential respondents to the survey were therefore representatives of the German investor firms only.

---

[76] The English version of the questionnaire is included in Appendix 7. We point out that the study is directed towards joint ventures with two partners: one German and one Chinese. Of course, there are exceptions to this, with three or more partners invested in one joint venture, but the study results will show that this is only true in very few cases and does not significantly undermine the assumptions made here.

Potential respondents were first identified by means of desk research or phone calls to companies that were known to be active in China. Questionnaires were then sent to them either by email or post.

The questionnaire was primarily designed to gather data for multiple regression analysis. This method allows testing the statistical significance of some derived hypotheses while controlling for a variety of general factors (in this case the factors discussed in Section III-2). For the gathering of data, many different question types were used, but a five-point Likert scale was most often used, enabling a respondent to indicate his degree of agreement (Hardy and Bryman 2004).[77]

In the following section, different approaches for measuring the *dependent variable* are presented.

1.2. Measurement Concepts for the Dependent Variable

The main objective of our study is to test for an empirical effect of a joint venture's cooperative constellation on the commitment by a German investor firm to transfer and share technologically sophisticated technology. The dependent variable should therefore be a measure representing the degree of sophistication that is transferred and/or shared.

As discussed in detail in Section II-1-2, the concept 'technological sophistication' has been referred to by a number of theoretical and empirical studies. Different ways to account for varying degrees of an organisation's technological sophistication have been to measure

1. the functional activities covered by an organisation
2. the technological resources with which an organisation is endowed
3. the technological capabilities of that operation, or
4. the performance of that organisation.

---

[77] The question format employed most often asked respondents for their opinion about a given statement on a five-point Likert scale, with 1 representing "I don't agree" and 5 representing "I agree".

However, few empirical studies exist that relate to technological sophistication as the dependent variable (Kabiraj and Marjit 1992), so that – at least from the author's point of view – most considerations discussed here are of an exploratory nature.

From the possibilities mentioned above, this study briefly refers to approach (1) by recording data on whether certain functional activities are covered by an organisation, such as 'research' or 'development'. Approach (4) is not appropriate for this study due to the large cross-sectional variety in product and process technologies.[78]

Approaches (2) and (3) are the main focus of this study. They are put into practice by referring to two previously developed scales described in Section II-1-2. The following paragraphs describe how scales representing the *technological resources* and *technological capabilities* of an organisation can be implemented to measure the technological sophistication of an international joint venture and – indirectly – the technology transfer commitment by a German investor firm.

1.2.1. Overview of Measuring Concepts

In the following paragraphs, we first describe how an *absolute* degree of technological sophistication is measured. We then derive measures for the *relative* degree technology transferred to an IJV by its German parent firm. Finally, we present a method to measure the relative commitment of a German investor firm to *share* its technological insights with a Chinese partner firm.

Let us assume that any German firm has some *maximum technological sophistication* useful for the production and marketing of a certain product or service. This knowledge could be created and/or nurtured at any location, for

---

[78] The approach of using performance figures as a proxy for technological sophistication is not well-suited for cross-sectional analysis of firms that operate in widely different industries as is the case in our study. They are usually employed for intra-firm analyses, technological performance over time, or theoretical models (see discussion in Section II-1-2)

example at the German headquarters (domestic) or a dedicated 'center for excellence'.[79] Irrespective of where this knowledge is located geographically, we assume that it can be accessed by the German headquarters and that it forms the maximum technological knowledge that a German investor firm can theoretically transfer to a foreign site.[80] We define the maximum technological knowledge embedded within the German firm and that is potentially transferable to the German-Chinese JV – at any cost – as $T(HQ)$.

The German-Chinese joint venture on the other hand, could be endowed with any level of technological sophistication, here referred to as $T(JV)$. Regarding the technological sophistication of the Chinese partner, we refer to the maximum endowment level as $T(P)$. Finally, the Chinese partner's degree of technological insight into the JV's operations is referred to as *PartnerInsight*. These variables can be used to produce relevant measures for dependent variables of this study.

These measures are:

- **T(JV):** The measure $T(JV)$ represents the absolute technological sophistication that actually 'ends up' in the Chinese joint venture. This absolute level is relevant because the technology deployed to China is put at risk of potential diffusion.
- **T(JV)/T(HQ):** The ratio $T(JV)/T(HQ)$ measures the *relative* degree to which the German firm endowed its foreign affiliate with technological sophistication, taking account of the absolute level of technological sophistication of the German headquarters. Based on the assumption that a JV's maximum technological sophistication is a result of

---

[79]   This depends on a firm's international R&D setup (v. Zedtwitz and Gassmann 2002). A Chinese-German joint venture can be one of the sites to create new knowledge, but R&D JVs are not the focus here. As pointed out by Beckmann and Fischer (1994, p. 635), if any new knowledge is created in a foreign site of an international organisation, it is usually country-specific application knowledge (resulting from country-specific application development) and is thus not suitable to be transferred to other country organisations.

[80]   Of course, the theoretical maximum knowledge embedded within a firm can never be perfectly duplicated. This is a result of the immobility of the necessary qualified personnel, the lack of resources or the result of cost-benefit considerations (Dong 2004).

technology transfer efforts by the German partner firm, the measure $T(JV)/T(HQ)$ can be used as a proxy for the 'relative technological investment' to the joint venture by its German partner.[81]

- **Shared technological insight:** Extending the considerations regarding the measure $T(JV)/T(HQ)$, one can measure the degree of technological insight that is shared by the German partner by controlling for the Chinese partner's actual understanding of the technologies employed by the joint venture. The corresponding measurement concept employed by this study can be expressed with the formula: *PartnerInsight* T(JV)/T(HQ)*.[82]

The following paragraphs will discuss the practical implementation of the above-discussed measurement concepts by using scales that express the sophistication of *technological resources* and *technological capabilities* (as theoretically discussed in Section II-1-2).

1.2.2. Measuring T(JV)

As outlined before, the measure *T(JV)* represents the absolute technological sophistication that actually 'ends up' in the Chinese joint venture. This level is relevant because the technology deployed to China is put at risk of potential diffusion. The following paragraphs will discuss the two approaches that this study uses to measure *T(JV)*, namely the scales for *technological resources* on the one hand and *technological capabilities* on the other.[83]

---

[81] This assumption will be discussed in detail when presenting the study results and in Chapter V.

[82] As the descriptive results in Section IV-2 will show, the Chinese partner's degree of insight is not correlated to the technological sophistication level of the joint venture. For measuring the actual degree of technological insight that the German investor shares with the Chinese partner, one therefore has to control for this actual degree of insight.

[83] The data gathered by this study would also allow the use of other measurement concepts for T(JV), such as the proportion of the final product that is manufactured in China. One straightforward way to measure T(JV) is also to ask survey respondents whether the technology that is employed by the joint venture is innovative or complex.

Technological Resources

The first approach to measure the sophistication of a joint venture's technological endowment is to refer to its endowment with *technological resources*. As noted by Sharif (1995, 1997) and elaborated on by Cohen (2004), these can be split up into the individual dimensions *technoware, inforware, humanware* and *orgaware* (see Section II-1-2). This study measures each dimension by asking respondents for the maximum level of technological resources employed by a joint venture with reference to a scale from 1 to 7 (see Appendix 2 and Appendix 7).

We compute correlations and Cronbach's alpha for the four dimensions *TechnowareJV, InforwareJV, HumanwareJV*, and *OrgawareJV* as measured for the survey data in the following table.[84]

This approach is simple and has been used by several authors before, e.g. Wilson (1997), Mansfield and Romeo (1980), Katz et al. (1996), and Weiss (1996). However, the insights provided by analysing these concepts do not significantly add to the insights provided here. They can be obtained by the author on request.

[84] Cronbach's alpha is the most common method of estimating the internal reliability of a scale. The usual formula used is alpha = k/k-1 + (1 − 1/v*w), where k is the number of items, v is the variance of the total score, and w is the sum of the variance of the items. If alpha comes out below 0.8, the reliability of the scale may need to be investigated further. See Hardy and Bryman (2004) for further reference.

Table 14: Information on Individual Dimensions of Technological Resources

Variables:	Description:
TechnowareJV	Max. level achieved by the JV with respect to **technoware** (scale from 1 to 7 - see Appendix)
InforwareJV	Max. level achieved by the JV with respect to **inforware** (scale from 1 to 7 - see Appendix)
HumanwareJV	Max. level achieved by the JV with respect to **humanware** (scale from 1 to 7 - see Appendix)
OrgawareJV	Max. level achieved by the JV with respect to **orgaware** (scale from 1 to 7 - see Appendix)

Descriptive Statistics:

	N	Min	Max	Mean	St- Dev.
TechnowareJV	34	2	7	4,65	1,48
InforwareJV	34	2	7	4,91	1,26
HumanwareJV	34	1	7	4,32	1,39
OrgawareJV	34	1	7	3,62	1,63

Correlations:

		Technoware JV	Inforware JV	Humanware JV	Orgaware JV
TechnowareJV	Correlation	1,00	0,19	,443(**)	,433(*)
	Sign.		0,27	0,01	0,01
InforwareJV	Correlation	0,19	1,00	0,29	0,10
	Sign.	0,27		0,09	0,57
HumanwareJV	Correlation	,443(**)	0,29	1,00	,471(**)
	Sign.	0,01	0,09		0,01
OrgawareJV	Correlation	,433(*)	0,10	,471(**)	1,00
	Sign.	0,01	0,57	0,01	

* Significant at the 5% level (two-tailed).
**Significant at the 1% level (two-tailed).

Cronbach's Alpha: 0,66   **without Inforware:** 0,71

Source: Survey results.

The statistics indicate that the average resource endowment per dimension differs and that some dimensions significantly correlate with each other. Because these three dimensions – technoware, humanware and orgaware – exhibit a rather high correlation one could argue in favour of a combined scale, but the Cronbach's alpha is below 0,8.[85] We therefore take account of the technological resources individually when describing study results. This is also reasonable due to the qualitative difference of each of these dimensions and our aim to derive differentiated results.

Technological Capabilities

The second alternative for measuring a JV's absolute level of technological endowment is a scale derived from the concept of *technological capabilities* (Ramanathan 1994). It distinguishes between *operative, acquisitive, innovative* and *supportive* technological capabilities (see Section II-1-2).

---

[85] As previously described, a Cronbach's alpha of below 0,8 suggests that the reliability of a scale is not guaranteed (Hardy and Bryman 2004).

We compute descriptive statistics, correlations and Cronbach's alpha for the four dimensions *OperativeJV*, *AcquisitiveJV*, *InnovativeJV*, and *SupportiveJV* as measured for the survey data in the following table.

Table 15: Information on Individual Dimensions of Technological Capabilities

Variables:	Description:
OperativeJV	Technological capabilities of JV with respect to **operative** capabilities (1=weak, 5=strong)
AcquisitiveJV	Technological capabilities of JV with respect to **acquisitive** capabilities (1=weak, 5=strong)
InnovativeJV	Technological capabilities of JV with respect to **innovative** capabilities (1=weak, 5=strong)
SupportiveJV	Technological capabilities of JV with respect to **supportive** capabilities (1=weak, 5=strong)

Descriptive Statistics:

	N	Min	Max	Mean	St- Dev.
OperativeJV	34	1	5	3,21	0,98
AcquisitiveJV	34	1	4	2,59	0,99
InnovativeJV	34	1	5	2,94	1,01
SupportiveJV	34	1	5	3,00	0,89

Correlations:

		Operative JV	Acquisitive JV	Innovative JV	Supportive JV
OperativeJV	Correlation	1,00	,592(**)	,594(**)	,733(**)
	Sign.		0,00	0,00	0,00
AcquisitiveJV	Correlation	,592(**)	1,00	,641(**)	,449(**)
	Sign.	0,00		0,00	0,01
InnovativeJV	Correlation	,594(**)	,641(**)	1,00	,539(**)
	Sign.	0,00	0,00		0,00
SupportiveJV	Correlation	,733(**)	,449(**)	,539(**)	1,00
	Sign.	0,00	0,01	0,00	

* Significant at the 5% level (two-tailed).
**Significant at the 1% level (two-tailed).

Cronbach's Alpha: 0,85

Source: Survey results.

The descriptive results suggest that all four dimensions of technological capabilities as observed in this sample are highly correlated. This is consistent with the argument by Saha and Nazrul (1998) that Ramanathan's classification is *multiplicative* in nature. Ramanathan (1994) himself draws an analogy with a 'stool with four legs and supports connecting the legs', meaning that all four capabilities as described by him are needed simultaneously.

This characteristic favours the use of a combined scale. We therefore produce a single scale *CapabilitiesJV*, with *CapabititiesJV* for any JV representing the arithmetic mean of the four dimensions *OperativeJV*, *AcquisitiveJV*, *InnovativeJV*, and *SupportiveJV*. A high Cronbach's Alpha for the reliability of a combined scale (0,85) supports the use of a combined scale.

Relationship between Technological Resources and Capabilities

The concepts 'technological resources' and 'technological capabilities' are closely related measures for the technological sophistication of an organisation. Variables measuring them as alternative dependent variables should therefore be expected to be correlated. The following table shows the correlations for the combined scale *CapabilitiesJV* with the individual dimensions of technological resources.

Table 16: All Correlations for Variables Expressing T(JV)

Variables:	Description:
TechnowareJV	Max. level achieved by the JV with respect to **technoware**
InforwareJV	Max. level achieved by the JV with respect to **inforware**
HumanwareJV	Max. level achieved by the JV with respect to **humanware**
OrgawareJV	Max. level achieved by the JV with respect to **orgaware**
CapabilitiesJV	**Scale** for level of **capabilities** (average of operative, acquisitive, innovative and supportive)

Correlations:

		TechnowareJV	InforwareJV	HumanwareJV	OrgawareJV	CapabilitiesJV
TechnowareJV	Correlation	1,00	0,19	,443(**)	,433(*)	0,31
	Sign.		0,27	0,01	0,01	0,07
InforwareJV	Correlation	0,19	1,00	0,29	0,10	,411(*)
	Sign.	0,27		0,09	0,57	0,02
HumanwareJV	Correlation	,443(**)	0,29	1,00	,471(**)	,746(**)
	Sign.	0,01	0,09		0,01	0,00
OrgawareJV	Correlation	,433(*)	0,10	,471(**)	1,00	,395(*)
	Sign.	0,01	0,57	0,01		0,02
CapabilitiesJV	Correlation	0,31	,411(*)	,746(**)	,395(*)	1,00
	Sign.	0,07	0,02	0,00	0,02	

* Significant at the 5% level (two-tailed).
**Significant at the 1% level (two-tailed).

Source: Survey results.

The results show that a joint venture's technological capabilities in this survey are most strongly correlated to the sophistication of the human resources it is endowed with. *InforwareJV* and *OrgawareJV* are also positively correlated to *CapabilitiesJV*, but not as strongly. This result seems intuitive as the technological capabilities of an organisation are derived from the technological skills of its people (see Section II-1-2).

For the analysis of study results, we will keep the five measurement dimensions *TechnowareJV*, *InforwareJV*, *HumanwareJV*, *OrgawareJV* and *CapabilitiesJV* apart, in spite of the correlations shown above. This allows for a differentiated analysis of each dimension.

1.2.3. Measuring T(JV)/T(HQ)

As outlined before, the concept *T(JV)/T(HQ)* is meant to represent the German firm's technological investment in the joint venture in China *relative* to the maximum technological level achieved by the German investor. It controls for the large difference in technological resources and capabilities by foreign investor firms themselves when assessing the commitment they show for their joint venture in China.

The smaller the difference between T(HQ) and T(JV), the higher we argue the technological investment of the German firm to be. Of course, this assumes that the technological endowment of the German-Chinese JV does indeed stem from the German partner.[86] We compute J(JV)/T(HQ) by *dividing* T(JV) by T(HQ), which is the most straightforward alternative. [87]

Technological Resources

The first candidate for a dependent variable relating to the relative investment in the JVs technological endowment is the ratio of technological resources between the German headquarters and the joint venture. Because the survey records the level for each dimension of technological resources for both the headquarters and the JV, a computation of differences is straightforward. Descriptive statistics for the variables are shown in the following table.

---

[86]   This assumption will be discussed in detail at a later state of this study (Sections IV-2-2 and V-1-2).

[87]   One could argue for other functions to express T(JV)/T(HQ). For example, the effort invested by a German firm to increase T(JV) could be argued to rise exponentially, because the same absolute difference of [T(HQ)-T(JV)] represents higher effort, the higher the absolute levels of T(HQ) and T(JV). For example, the scale used for technoware in the survey (Appendix 3 and Appendix 7) assigns the value '1' to manual tools (e.g. screwdriver, hand drill), '2' to powered equipment (e.g. grinder, power drill) and '3' to general purpose facilities (e.g. milling machine, lathe). One could argue that the effort required to change from 2 to 3 is much higher than from 1 to 2. However, we do not employ this approach in order to avoid potential bias or excessive complexity.

Table 17: T(JV)/ T(HQ) for Technological Resources as Dependent Variable

Variables:	Description:
TechnowareRel	T(JV) / T(HQ) with respect to **technoware**
InforwareRel	T(JV) / T(HQ) with respect to **inforware**
HumanwareRel	T(JV) / T(HQ) with respect to **humanware**
OrgawareRel	T(JV) / T(HQ) with respect to **orgaware**

Descriptive Statistics:

	N	Min	Max	Mean	St- Dev.
TechnowareRel	34	0,29	1,00	0,77	0,21
InforwareRel	34	0,29	1,00	0,75	0,18
HumanwareRel	34	0,17	1,00	0,65	0,20
OrgawareRel	34	0,14	1,00	0,60	0,21

Correlations:

		TechnowareRel	InforwareRel	HumanwareRel	OrgawareRel
TechnowareRel	Correlation	1,00	0,00	0,26	0,20
	Sign.		0,98	0,14	0,26
InforwareRel	Correlation	0,00	1,00	,344(*)	0,26
	Sign.	0,98		0,05	0,13
HumanwareRel	Correlation	0,26	,344(*)	1,00	0,22
	Sign.	0,14	0,05		0,21
OrgawareRel	Correlation	0,20	0,26	0,22	1,00
	Sign.	0,26	0,13	0,21	

* Significant at the 5% level (two-tailed).
**Significant at the 1% level (two-tailed).

Cronbach's Alpha for scale:	0,52

Source: Survey results.

As the table illustrates, the technological resources of the joint venture on average reach between 60% (Orgaware) and 77% (Technoware) of the headquarters' level. There is no case in our sample for which the technological resources of the joint venture exceed the resources of the German firm, but there are cases in which the JV's level of technological resources equals the maximum level of any site achieved by the German company worldwide.[88]

For deriving the relative degree of technological resources transferred, the data again suggests investigating each of the dimensions technoware, inforware, humanware and orgaware separately, because only *HumanwareRel* and *InforwareRel* are positively correlated to a significant degree.

---

[88] This should be the case by definition of the variables, because the maximum technological level of the German firm worldwide should include the maximum level of the joint venture at hand. It should be noted here that there are a few cases in which the Chinese JV *partner* has higher resource endowment than the JV. These cases are shown in Section IV-2-2.

Technological Capabilities

The next table shows descriptive information and correlations to the different dimensions of technological capabilities when expressed as relative differences between the HQ and the JV. It also computes Cronbach's alpha for a possible combined scale.

Table 18: T(JV)/T(HQ) for Technological Capabilities as Dependent Variable

Variables:	Description:
OperativeRel	T(JV) / T(HQ) with respect to **operative capabilities**
AcquisitiveRel	T(JV) / T(HQ) with respect to **acquisitive capabilities**
InnovativeRel	T(JV) / T(HQ) with respect to **innovative capabilities**
SupportiveRel	T(JV) / T(HQ) with respect to **supportive capabilities**

Descriptive Statistics:

	N	Min	Max	Mean	St- Dev.
OperativeRel	34	0,25	1,00	0,73	0,29
AcquisitiveRel	34	0,20	1,00	0,63	0,46
InnovativeRel	34	0,20	1,00	0,68	0,30
SupportiveRel	34	0,25	1,00	0,67	0,23

Correlations:

		OperativeRel	AcquisitiveRel	InnovativeRel	SupportiveRel
OperativeRel	Correlation	1,00	,844(**)	,802(**)	,794(**)
	Sign.		0,00	0,00	0,00
AcquisitiveRel	Correlation	,844(**)	1,00	,876(**)	,720(**)
	Sign.	0,00		0,00	0,00
InnovativeRel	Correlation	,802(**)	,876(**)	1,00	,761(**)
	Sign.	0,00	0,00		0,00
SupportiveRel	Correlation	,794(**)	,720(**)	,761(**)	1,00
	Sign.	0,00	0,00	0,00	

* Significant at the 5% level (two-tailed).
**Significant at the 1% level (two-tailed).

Cronbach's Alpha for scale:	0,92

Source: Survey results.

The data shows that in the case of technological capabilities, the average relative endowment ranges from 63% (*AcquisitiveRel*) to 73% (*OperativeRel*) of the German headquarters. Contrary to the case of technological resources, the individual dimensions of technological capabilities are all significantly correlated. A high Cronbach's alpha (0,92) suggests the use of a combined scale that expresses the relative difference between the technological capabilities of the HQ and the joint venture. This scale will be named *CapabilitiesRel*.

Relationship between Relative Technological Resources and Capabilities

The following table shows correlations for all concepts discussed above. The technological capabilities are already expressed as a scale (*CapabilitiesRel*). Contrary to that, the individual dimensions that represent technological resources remain separated.

Table 19: All Correlations for Variables Expressing T(JV)/T(HQ)

Variables:	Description:
TechnowareRel	T(JV) / T(HQ) with respect to **technoware**
InforwareRel	T(JV) / T(HQ) with respect to **inforware**
HumanwareRel	T(JV) / T(HQ) with respect to **humanware**
OrgawareRel	T(JV) / T(HQ) with respect to **orgaware**
CapabilitiesRel	T(JV) / T(HQ) scale for **capabilities** (mean of operative, acquisitive, innovative and supportive)

Korrelationen

		TechnowareRel	InforwareRel	HumanwareRel	OrgawareRel	CapabilitiesRel
TechnowareRel	Correlation	1,00	0,00	0,26	0,20	0,02
	Sign.		0,98	0,14	0,26	0,91
InforwareRel	Correlation	0,00	1,00	,344(*)	0,26	,438(**)
	Sign.	0,98		0,05	0,13	0,01
HumanwareRel	Correlation	0,26	,344(*)	1,00	0,22	**0,32**
	Sign.	0,14	0,05		0,21	0,07
OrgawareRel	Correlation	0,20	0,26	0,22	1,00	0,30
	Sign.	0,26	0,13	0,21		0,08
CapabilitiesRel	Correlation	0,02	,438(**)	**0,32**	0,30	1,00
	Sign.	0,91	0,01	0,07	0,08	

* Significant at the 5% level (two-tailed).
**Significant at the 1% level (two-tailed).

Source: Survey results.

The table shows that *CapabilitiesRel* is significantly correlated with *InforwareRel* (0,44**) and *HumanwareRel* (0,32). This compares well to previous computation of correlations regarding the absolute level *T(JV)*, where we found correlations for the same two dimensions.

This result seems very intuitive given the definitions of technological capabilities and the two dimensions of technological resources, because they both relate to the degree of technological insight of staff working at the JV.

In the next section, we will extend the analysis by the actual degree of insight the Chinese joint venture partner has.

1.2.4. Measuring Shared Technological Insight

The previous section described the measure $T(JV)/T(HQ)$ as a potential dependent variable measuring the relative degree of technological investment that the German investor commits to (given that some assumptions hold). However, this does not take into account to which degree the Chinese partner actually understands the technology employed by the joint venture.

We thus introduce a measurement concept that is expressed as *PartnerInsight* * $T(JV)/T(HQ)$, where *PartnerInsight* represents the Chinese partner's degree of insight. The variable that represents *PartnerInsight* is measured as the arithmetic mean of two other variables. These variables and descriptive information about them are computed in the table below.

Table 20: Descriptive Statistics for *PartnerInsight*

Variables:	Description:				
PartnerUnderst	The Chinese partner understands all of the technologies employed by the JV				
PartnerProduc	The Chinese partner could produce the JV's products by himself				

Descriptives:

	N	Min	Max	Mean	St- Dev.
PartnerUnderst	33	1	5	2,58	1,20
PartnerProduc	33	1	5	2,67	1,49

Scale Indicators:

Correlation:	0,72**	** significant at 0,01 (two-tailed)
Cronbach's Alpha:	0,83	

Source: Survey results. Scales from 1= "don't agree" to 5 = "I agree".

The descriptive statistics show that the two variables employed are highly correlated and can be used as a combined scale (*PartnerInsight*). Moreover, the actual degree of insight by the partner varies from the minimum possible value (1) to the maximum possible value (5).

It is thus necessary to control for this insight when investigating the degree of technology shared by a German investor with the Chinese partner.[89] As we will show in the descriptive results section, *PartnerInsight* is not correlated with the individual dimensions for measuring $T(JV)/T(HQ)$, indicating that the questions

---

[89]　Again, the degree of insight only correctly measures the degree of sharing by the German investor if the technological assets and capabilities of the German investor exceed those of the Chinese partner.

(a) whether a JV has a high technological endowment and whether (b) the German investor shares technological insights with the Chinese JV partner are two different questions. For measuring the actual degree of technological insight that the German investor *shares* with the Chinese partner, we therefore propose the measurement concept *PartnerInsight* T(JV)/T(HQ)* as the appropriate one.

We next show descriptive statistics and correlations for proposed dependent variables that result from multiplying *PartnerInsight* with each of the dimensions *TechnowareRel*, *InforwareRel*, *HumanwareRel*, *OrgawareRel*, and *CapabilitiesRel* (Table 21). Because the resulting variables are supposed to indicate the relative degree of knowledge sharing by the German investor with the Chinese partner, we label these variables *Shared_Technoware*, *Shared_Inforware*, *Shared_Humanware*, *Shared_Orgaware*, and *Shared_Capabilities*.

Table 21: Descriptive Statistics for Indicators Regarding Shared Technological Insight

Variables:	Description:
Shared_Technoware	Shared knowledge with JV Partner with respect to **technoware**
Shared_Inforware	Shared knowledge with JV Partner with respect to **inforware**
Shared_Humanware	Shared knowledge with JV Partner with respect to **humanware**
Shared_Orgaware	Shared knowledge with JV Partner with respect to **orgaware**
Shared_Capabilities	Shared knowledge with JV Partner with respect to **capabilities**

Descriptive Statistics:

	N	Min	Max	Mean	St- Dev.
Shared_Technoware	33	0,29	4,00	2,07	1,14
Shared_Inforware	33	0,40	4,50	2,00	1,15
Shared_Humanware	33	0,42	4,00	1,68	0,97
Shared_Orgaware	33	0,38	3,50	1,50	0,84
Shared_Capabilities	33	0,44	5,00	1,79	1,49

Correlations:

		Shared_ Technoware	Shared_ Inforware	Shared_ Humanware	Shared_ Orgaware	Shared_ Capabilities
Shared_Technoware	Correlation	1,00	,850(**)	,790(**)	,728(**)	,551(**)
	Sign.		0,00	0,00	0,00	0,00
Shared_Inforware	Correlation	,850(**)	1,00	,783(**)	,704(**)	,753(**)
	Sign.	0,00		0,00	0,00	0,00
Shared_Humanware	Correlation	,790(**)	,783(**)	1,00	,586(**)	,605(**)
	Sign.	0,00	0,00		0,00	0,00
Shared_Orgaware	Correlation	,728(**)	,704(**)	,586(**)	1,00	,649(**)
	Sign.	0,00	0,00	0,00		0,00
Shared_Capabilities	Correlation	,551(**)	,753(**)	,605(**)	,649(**)	1,00
	Sign.	0,00	0,00	0,00	0,00	

* Significant at the 5% level (two-tailed).
**Significant at the 1% level (two-tailed).

Source: Survey results.

The table above indicates that on average, the shared technological insight is between 1,5 (*Shared_Orgaware*) and 2,04 (*Shared_Technoware*). This logically follows from the combination of *PartnerInsight* (average around 2,62) and the individual scales for T(JV)/T(HQ) (average around 70%). The high correlations between the individual scales follow from multiplying each individual scale for *T(JV)/T(HQ)* with *PartnerInsight*. Although the correlations are high, we argue that each indicator measures a different concept, which is why we will investigate results separately for each indicator.

The next section will briefly discuss other possible concepts for measuring the technology transfer efforts by the German investor firm that have not been implemented. This discussion serves to reflect on other possible measures and how they would be related to the measures we employ. Section IV-2 then presents the survey results.

1.2.4. Possible Measures of Technology Transfer that are Not Pursued

Several alternative measures to represent the technology transfer efforts undertaken by the German investor firm could have been employed but are neglected here to maintain this study's focus.[90]

As we described in detail in Section II-2-3, one can argue that a technology's characteristics influence the way in which international firms transfer it to foreign markets. For example, old and established technology is less costly to transfer (Teece 1977), whereas new and profitable technology is unlikely and difficult to be transferred via a JV (Mansfield 1979). One could turn this argument around and claim that old or previously transferred technology in a given joint venture represents a low technological commitment, whereas new and advanced technology represents a high commitment by the technology sender. One variable that one could take into account is the question of how advanced the product in China is compared to world standards.[91] One could also record whether the transferred technology is a core competence of the German firm.

Secondly, one could measure the absolute level of technological sophistication by measuring the share of the final product that the joint venture is able to produce to be sold in China. The argument is that the more of a given product a joint venture can produce, the higher its technological capability. We have recorded this information – distinguishing between the weight and the value

---

[90]   This discussion is somewhat related to the discussion of general factors in Section III-2, where we summarise the characteristics of technology that can influence the final result of technology transfer. We therefore keep the theoretical discussions short and refer the reader to the above-mentioned section.

[91]   This idea has been included in the survey as the questionnaire question: "The described technology that has been transferred by us to the JV is modern when compared to our most modern plants in the world" (scale from 1 to 5: 1= I don't agree, 5 = I agree).

share that the joint venture produces[92] – and give some descriptive results about it in Section IV-2-1.

Finally, one could simply ask survey respondents whether the technology employed by the joint venture is innovative and/or complex. It would be easy to use and one could argue that a joint venture employing technology that is neither innovative nor complex has a low technological endowment, whereas JVs using complex and innovative technologies have a high technological endowment. However, this measure has the drawback of relying too much on the subjective assessment of the respondent for the purposes of this study. Another drawback is the low comparability of respondents' opinions across the sample. What one finds innovative or complex depends to a large extent on personal experiences or benchmarks.

The survey data recorded in the course of this study allows the analysis of the data referred to here. However, we do not pursue these approaches for the reasons explained above and for the sake of clarity.

We next present the empirical research results.

2. Empirical Research Results

The following sections will present the results of this study. We first give an overview of the course of the investigation and the resulting sample. Section IV-2-2 contains some descriptive results that are related to the topic of technology transfer and a JVs technology endowment. Section IV-2-3 presents specific statistical results that are used to empirically evaluate the hypotheses presented in the previous chapter. The findings in this chapter are then summarised and discussed in Section IV-3.

---

[92]   This is because one could think of the 'common' case where the local JV produces most of the parts of a product, but the key components with high value are shipped from Germany (see recommendations by the consulting firm Roland Berger with regard to IP protection in China (von Keller et al. 2005).

The presented results provide interesting and new insights into the topic of cooperation and technology endowment in international joint ventures and also present a new application of a game theoretic model on inter-company cooperation.

2.1. Course of the Investigation and Sample Description

As previously described, the empirical research for this study was implemented in two steps. In November 2003, structured interviews were conducted in order to provide first insights into the topic. Key concepts were discussed with a total of nine different expert or expert teams with very different backgrounds – industry practitioners, (publicly funded) consultants, academics and one lawyer. Short descriptions of the interviews conducted are provided in Appendix 3.

Based on the insights from these expert interviews, a questionnaire survey among German industrial firms with current investments in international joint ventures in China was conducted. This survey took place between February 1st and August 31st 2007. From the questionnaires that were sent out, 34 were received back before the end of the survey period. All were suitable for analysis.[93] This translates into a response rate of about 13%, which is not high, but can be rated as acceptable for a company survey targeting high-level managers. [94]

The following paragraphs will give descriptive information on the sample obtained. We first elaborate on the participating firms, then on the respondents, and finally on the described joint ventures. To start out with,

Figure 26 shows the participating firms according to their number of employees.

---

[93] This is also due to the fact that in case a question was not (or not correctly) filled in, the author contacted the respondent and was able to gather all necessary information.
[94] As previously described, 259 joint ventures were identified and confirmed as currently active international joint ventures of German firms in China.

Figure 26: Participating Firms - Number of Employees

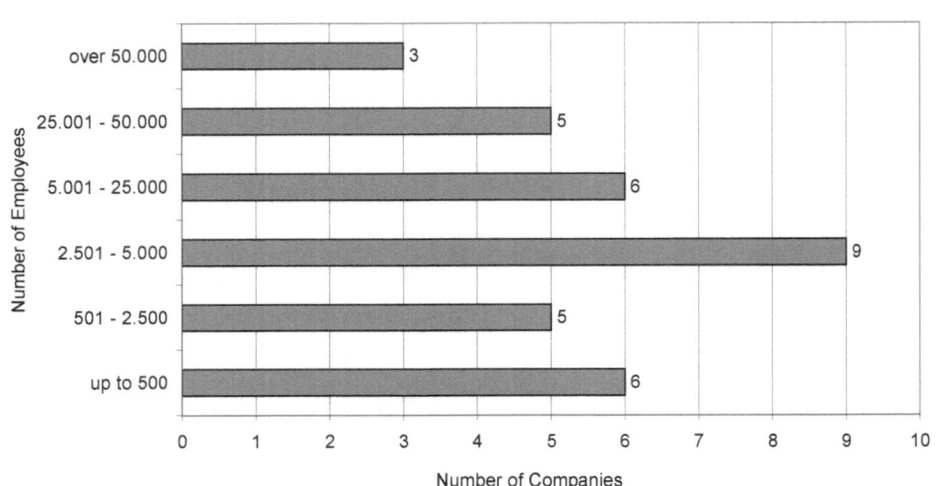

Source: Survey results.

As the figure above shows, companies of all sizes have participated in the survey. Three participants represent firms with over 50.000 employees worldwide. On the other hand, there are also five participating firms with fewer than 500 employees. The global sales figures of the participating companies give a very similar picture, ranging from under EUR 5 million to over EUR 100 billion per year.[95]

Next, Figure 27 gives the personal backgrounds of the respondents. It shows that the persons who are knowledgeable about the history and the current status of a certain joint venture in China and the particular topic of technology endowment are mostly high-ranking managers, both within headquarters (board member, CEO, director of a business division) or the respective joint venture (General Manager, CFO).

---

[95]   The sales levels are not shown here with a separate figure, as they will appear in figures shown at a later stage.

Figure 27: Background of Respondents - Positions

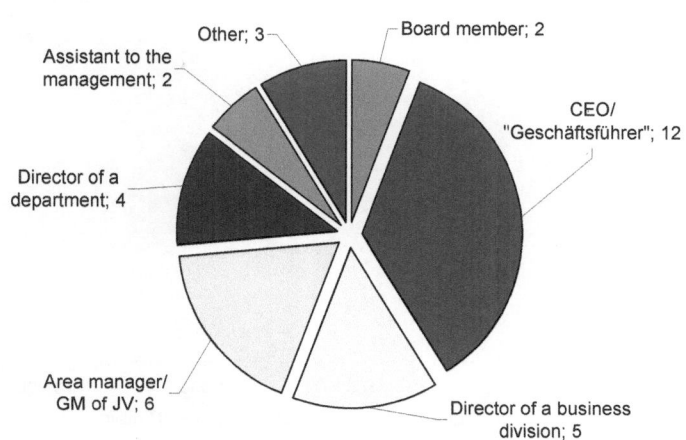

Source: Survey results.

Of the 34 managers shown in the above figure, 21 are geographically located in the German firms' headquarters, 11 in China, and two in other locations. Out of the 11 persons located in China, eight respondents are the respective joint venture's General Manager, whereas the other three persons hold other high-level positions such as 'CFO of the joint venture'.

The next set of figures relates to basic information on the joint ventures that are represented in this study. The figure below shows the joint venture's sales levels in combination with the German firm's worldwide turnover. As could be expected, the JV turnover levels tend to correlate with the worldwide turnover of the German investor firm (correlation of +0,23). Based on the data shown below, one can also compute the ratio of JV sales to the German firms' worldwide turnover. This ratio ranges between 0,01% and 35%, with an average level of around 5%.

Figure 28: Background of Joint Ventures - Turnover

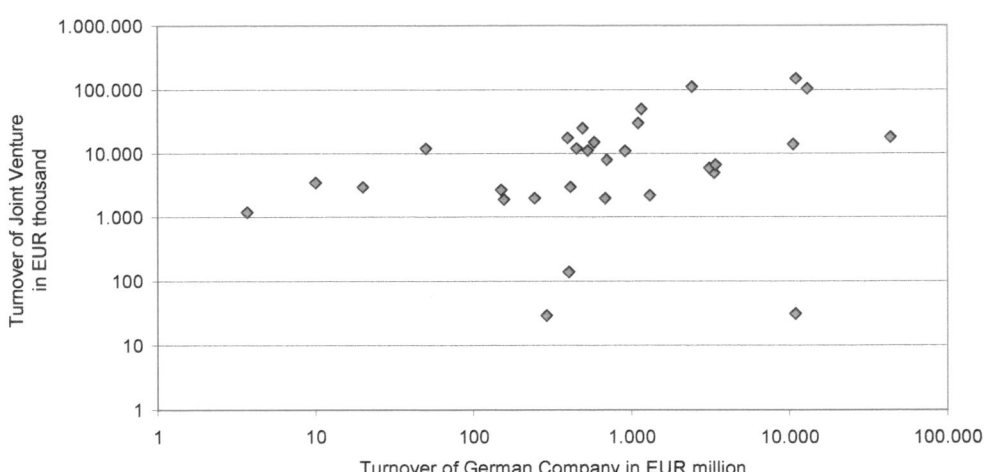

Source: Survey results (scales are logarithmic).

Other relevant dimensions for describing JVs are their age and the investor's ownership share. As the figure below shows, the JVs in this sample are between 3 months and 267 months (over 22 years) old. The ownership shares of the survey respondents range between 12,5% and 99% and show a positive, but insignificant correlation to a JV's age (+0,13).[96]

---

[96] The reader might conclude from this data that the foreign ownership share in a joint venture does not increase over time, at least not to a significant degree. This question is not easy to answer, however. On the one hand, individual case studies investigated by the author do confirm the argument that a lot of 'old' JVs exist in which both partners are satisfied with a certain ownership division. On the other hand, the data above represents only the joint ventures that *are* still joint ventures: many of the joint ventures listed in the AHK Database from 2004 have been turned into a whole subsidiary until September 2007 and thus are not part of this sample.

Figure 29: Background of Joint Ventures – Age of JV and Ownership Share

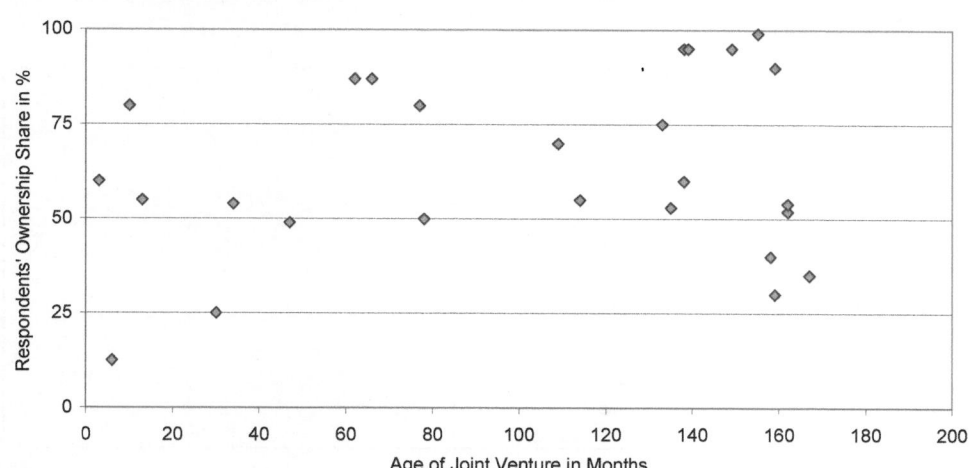

Source: Survey results.

The next figure shows the joint ventures' contract duration and the amount of registered capital. While the registered capital of the sample JVs ranges from EUR 1 million to EUR 100 million, the duration of the JV contracts ranges from 10 years to 99 years.[97]

---

[97]   99 years was used by the respondents to represent 'indefinite' contract duration. There are less than 34 dots in the figure because the registered capital was not revealed by all firms.

Figure 30: Background of Joint Ventures – Capital Investments and Contract Duration

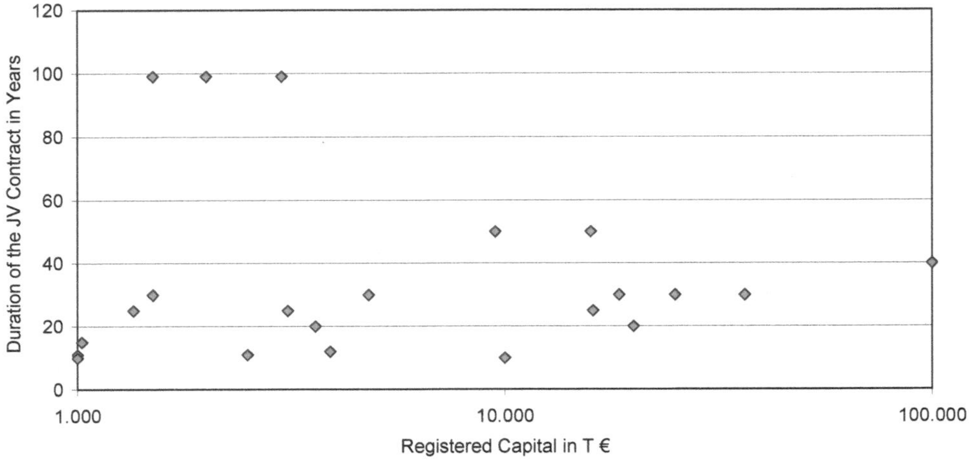

Source: Survey results.

A significantly positive relationship between these two JV characteristics could not be proven (correlation of +0,02). German firms therefore do not seem to insist on longer JV contract durations as the total amount of registered capital increases.

Finally, the JVs in this sample have been characterised by respondents by assigning them to joint venture types. The typology used here is based on the results by Hermann (1988) presented earlier and was extended by the category 'production JV' (see Fu 2005).[98]

---

[98]   Because multiple answers were given, the total sum of answers exceeds the number of cases.

Figure 31: Background of Joint Ventures – Joint Venture Types

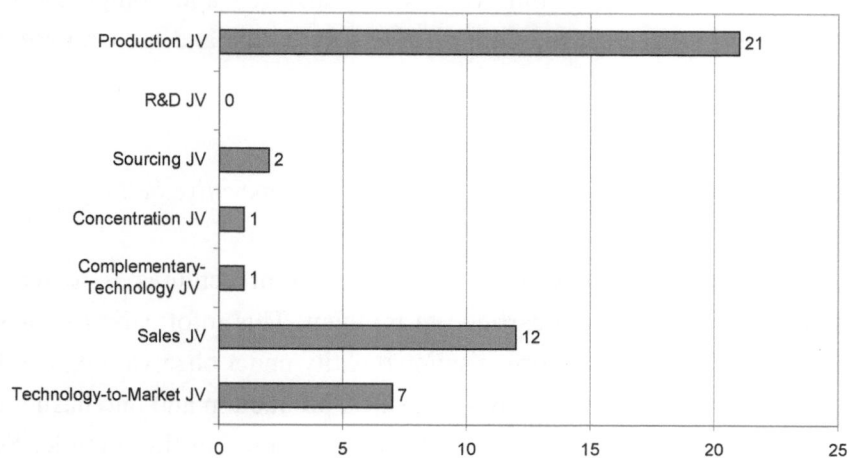

Source: Survey results.

The data shows that the joint ventures in this sample overwhelmingly carry out functions related to production and sales. Firms predominantly pursue market entry accompanied by local production or use the Chinese JV as a low-cost manufacturing site. One related and relevant category is 'Technology-to-Market JV' with seven answers. The other types are not well-represented or are not applicable at all (no R&D JVs).

In the following sections, we will provide descriptive results of this study that focus on the basic characteristics of the joint ventures under observation, their cooperative setting, the German investor's technology transfer behaviour and the resulting JV's technological endowment.

2.2. Descriptive Results

The descriptive results of this study focus on the analysis of the technology employed by joint ventures, the characterisation of the technology transfer effort by the foreign investor firm and the resulting technology endowment of an international joint venture. To start out with, some basic characteristics of the observed joint ventures are described, focusing on the cooperative setting and the division of tasks within the partnership. Second, the technological

167

sophistication of the joint ventures' output is investigated. Finally, the technology endowment of joint ventures is analysed and compared with the technology endowment of the German firm worldwide and the Chinese partner firm.

2.2.1. Characteristics of Joint Ventures and the Cooperative Setting

One previously described way to characterise joint ventures is to record the functional activities that are carried out by them. This information allows basic insights into the skills of the organisational entity under observation (see Section II-1-2). The following figure shows that sales, production and purchasing are the functions most often carried out by the joint ventures in this sample. Service, marketing and development are functions carried out by more than half of the JVs. Only four out of 34 joint ventures cover the research function.

As discussed earlier, the presence of certain functional responsibilities, especially the development and the research function within an organisational unit are an indicator for technological sophistication.

Figure 32: Functional Activities Carried Out by Joint Ventures

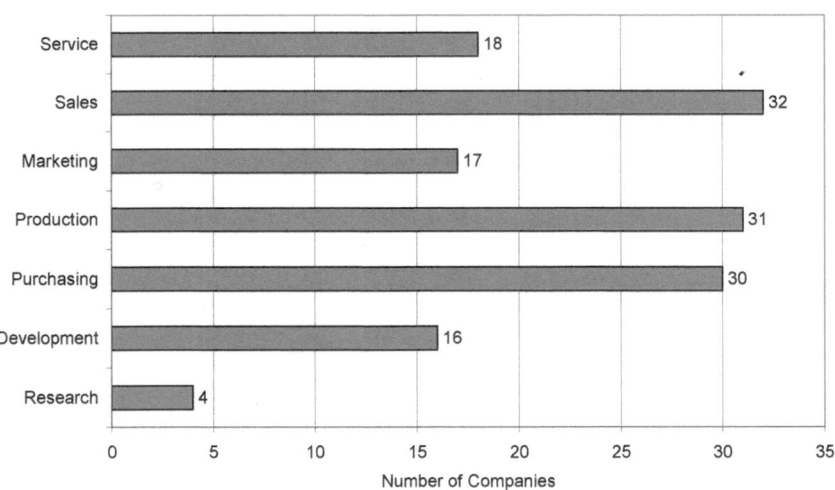

Source: Survey results.

As the next figure shows, the research and the development function are usually carried out under the sole responsibility of the German partner. A qualitative description of who does what in every joint venture is provided in Appendix 6.

Figure 33: Relative Responsibility for Functional Activities in Joint Ventures

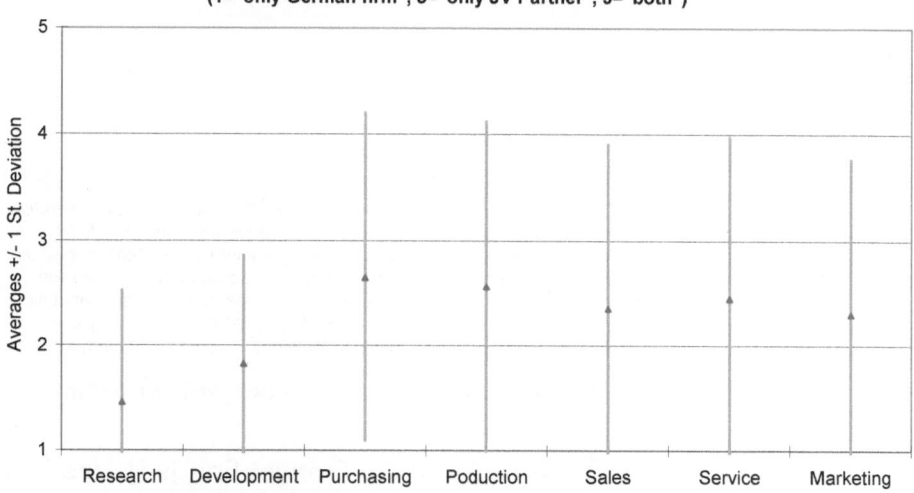

Source: Survey results. The total length of each line represents two standard deviations.

For further investigating the cooperative setting of the joint ventures, the next figure shows summary results for a set of questions that relate to the nature of interaction within the relationship and the levels of respect, trust, and friendship between the employees of the JVs.

Figure 34: Results regarding the Nature of Interaction between Partners

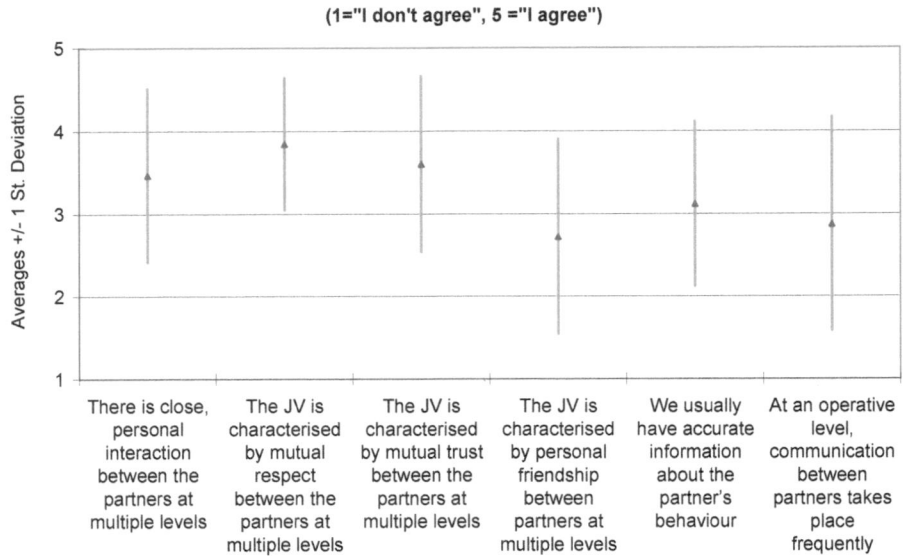

Source: Survey results. The total length of each line represents two standard deviations.

The respondents that answer on behalf of the German firm judge the level of respect and trust within their JVs to be high. On the other hand, they don't think that JVs can be characterised by friendship. The average opinion on the frequency of inter-partner communication and the accuracy of information received by the German partner is rather neutral.

An important characteristic regarding inter-alliance cooperation that is registered by many researchers (Weiss 1996, Fu 2005) is the question of whether a given market entry form represents the investor's first choice or not. In our case, 22 out of 34 JVs represent a 'free choice' of the German investor. In 11 cases, the German investor would have preferred to use a different market entry form (usually a WFOE) but was forced to use a joint venture due to investment regulations.[99]

---

[99]   Appendix 6 provides an overview of which JVs were result of free entry choice and which ones were not.

Table 22: Background of Joint Ventures – Preferred Market Entry Choice?

JV was preferred market entry choice:	22
The preferred choice would have been	
- WFOE:	11
- Licensing:	1
Total	34

Source: Survey results.

In the cases where German investors were 'forced' to use joint ventures, one could expect a different nature of interaction between partners or technology endowment behaviour by the German investors. This question will be answered in later sections. First, however, we go on to present descriptive results regarding the technological characteristics of the JVs output.

2.2.2. Technological Sophistication of Joint Ventures

In this section, we characterise the technology used by the joint ventures in this sample. This information can be used to derive the technology endowment commitment by German investors. We start out by computing the types of manufacturing processes used for all products in this sample, using the typology proposed by Woodward (1965).

Figure 35: Production Process Types

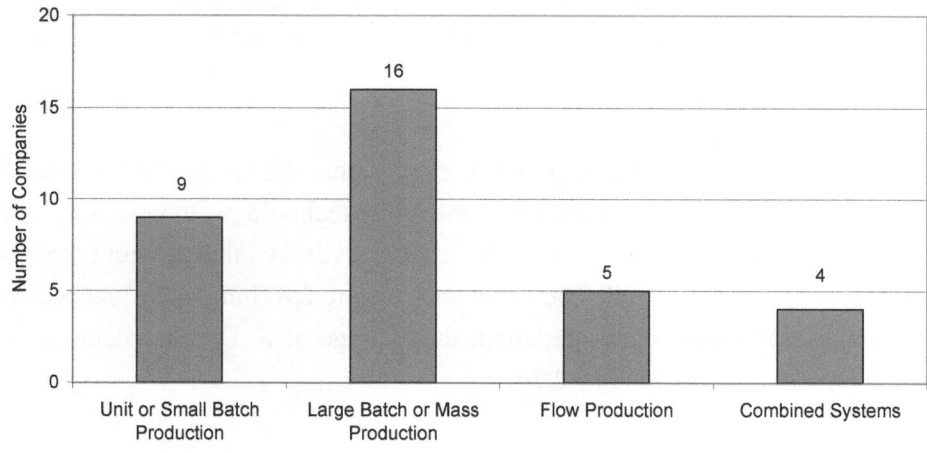

Source: Survey results.

As the figure shows, the majority of JVs in this sample employ unit or small batch (9) or large batch or mass production (16) technology. Only 5 JVs operate in industries that employ flow production (e.g. chemicals or pharmaceuticals) and only 4 JVs use combined systems.

A first indicator of the technological sophistication is the respondents' subjective evaluation on whether the technology employed by the joint venture is innovative and/or complex. Organisations that are able to use innovative or complex technology have arguably a rather high technological sophistication. The next figure shows the frequency of answers given for both concepts.

Figure 36: How Innovative and Complex is the Technology Employed?

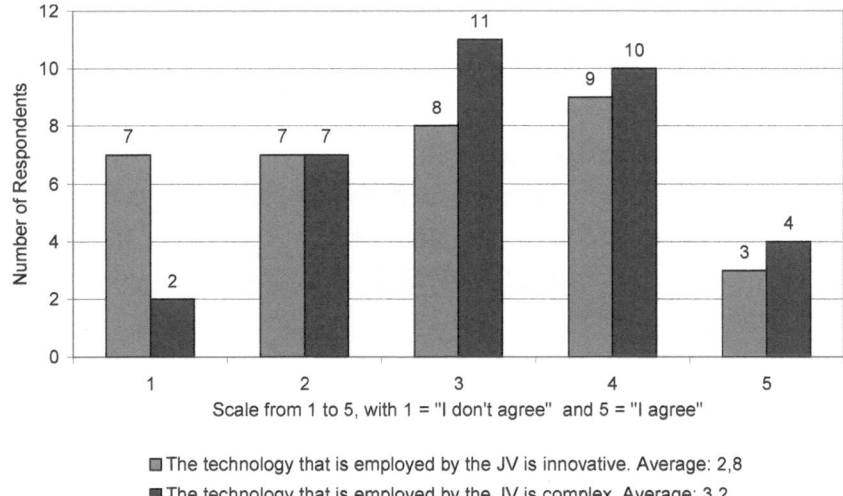

Source: Survey results.

The results show that there is quite a variety within the sample. It includes JVs that are evaluated as employing very innovative technology (three cases) and/or very complex technology (four cases), but also seven JVs that are not employing innovative technology at all. The mean answers are 2,8 (innovative) respectively 3,2 (complex), which represents about the average of 3. Both concepts show a strong positive correlation of +0,76.

Another indicator of a JVs technological sophistication is the share of the final output that is actually contributed by the joint venture. This share can be computed both with regard to the value of the final output and the physical weight of the final product.[100] As the next figure shows, these two measuring concepts are strongly correlated to each other, but are not exactly the same.

Figure 37: Share of Value Creation by Joint Ventures

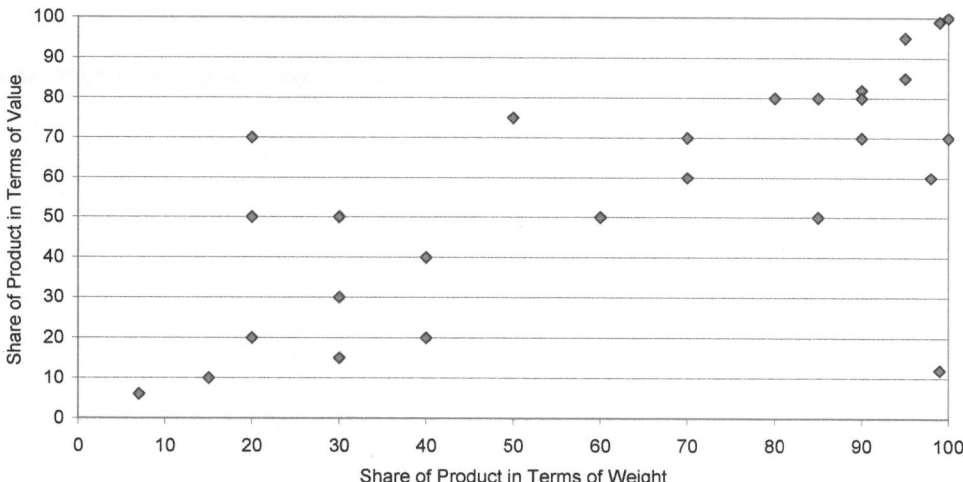

Source: Survey results.

As the figure above shows, the JVs in this sample vary a lot regarding their contribution to the final product: some JVs contribute less than 20% of value added, while some JVs reach a local content rate of almost 100%. A high correlation of both measures (+0,74) confirms the intuition that they are inextricably linked. However, the local content rate when measured in value tends to be lower than the local content measured in physical weight: the average contribution with respect to the product's final weight is 68%, whereas the average with respect to the final value of the product is 62%. This observation fits to some quotes of managers stated in previous sections that

---

[100] This indicator has been derived from expert interviews. The mentioned share relates to one single product that respondents were asked to describe, not to the average across all of a JV's products.

foreign firms tend to import high value components while manufacturing bulky components locally.

A final measure for the technological sophistication of a product is the comparison of product life cycle stages in Europe and China. As the next figure shows, the products manufactured and sold by the JVs are by European standards - mostly mature products or even products in saturated markets. In the context of the Chinese market, these products are evaluated to be in the growth stage of the product life cycle.

Figure 38: Number of Products Produced by Life Cycle Stage in Europe and China

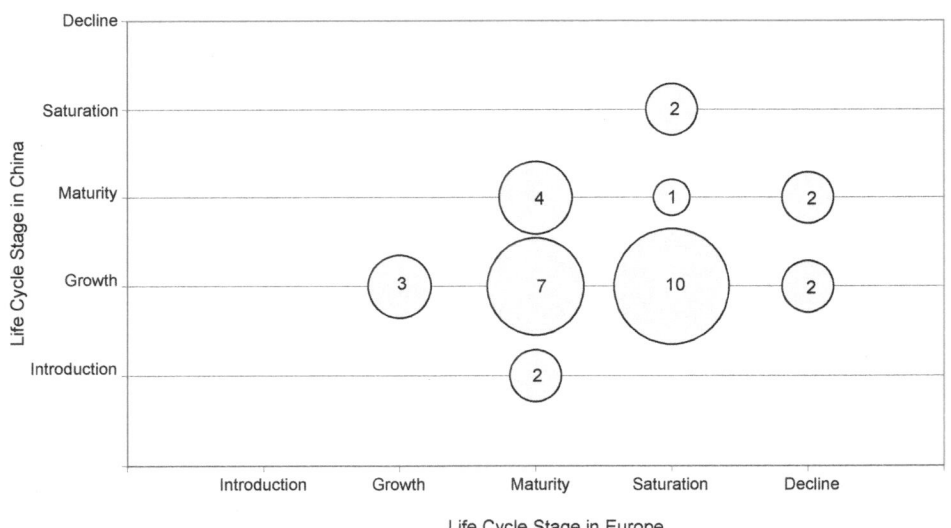

Source: Survey results.

One can summarise the results of this section with the observation that the joint ventures in this sample are endowed with sufficient technological resources and capabilities to manufacture a significant portion of the final product. However, these final products represent mature products when measured by European standards. The production technology used can thus be rather complex at times, but is evaluated to be innovative only in a few cases. Finally, the relative responsibility regarding all functions tends to be with the German partner.

Next, we present descriptive findings on the technology endowment of joint ventures when measured by some of the scales discussed in Section IV-1-2.

2.2.3. Technology Endowment of Joint Ventures

One main pillar of this survey is the detailed measurement of a joint venture's technology endowment. For this measurement, we implement the concepts discussed in Section IV-1-2 regarding technological sophistication and technological capabilities. In the following paragraphs, we will present some descriptive findings regarding the 'technology profile' of joint ventures, taking into account the maximum sophistication level of the German company (HQ), the joint venture (JV), and the Chinese JV partner (P).
We first relate to the four dimensions of technological resources, namely technoware, inforware, humanware and orgaware and their relationship to one another. The following figure shows the average resource endowment levels (red dot) as well as the standard deviations (green lines) indicated across the sample.[101]

Figure 39: Average Levels of Technological Resources for HQ, JV and the Partner

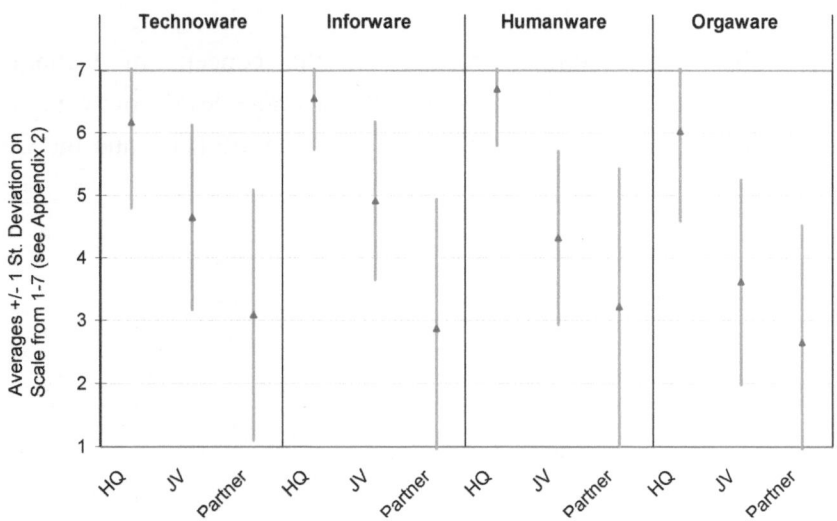

Source: Survey results.

---

[101]   The total length of the green line represents two standard deviations.

As this figure indicates, the 'typical' profile of a JV partnership regarding the sophistication of technological assets is that the German firm exhibits the highest resource endowment, followed by the joint venture itself. The Chinese partner, on the other hand, lags behind.[102]

The absolute levels of the joint venture's endowment regarding the four different dimensions represent the implementation of the measurement concepts of T(JV) discussed earlier. Because the descriptive results of *TechnowareJV*, *InforwareJV*, *HumanwareJV* and *OrgawareJV* are already depicted in Table 14, we don't reproduce them again here. However, it is worthwhile repeating that the absolute levels of the JVs' technoware, humanware, and orgaware are positively correlated to a significant degree, but that the correlation is not high enough to argue in favour of a combined scale that represents a joint venture's technological resources. *InforwareJV* is not correlated to the other three dimensions at all.

Different results hold for the measurement concept T(JV)/T(HQ). Here, only the dimensions *HuanwareRel* and *InforwareRel* are significantly correlated to each other (see Table 15). This indicates that the absolute level of a joint venture's technological endowment represents a slightly different concept than its relative endowment when taking into account the maximum sophistication level achieved by the German headquarters.

Next, we show corresponding results for the concept of technological *capabilities*. The following figure shows the average levels of technological capabilities for the German firm (HQ), the joint venture (JV), and the Chinese partner (P).

---

[102] These are only average results: there exist cases in which the Chinese partner's resource level either exceeds the JV or even the German firm (to be summarised in Table 24 in Section IV-2-3).

Figure 40: Average Levels of Technological Capabilities for HQ, JV and the Partner

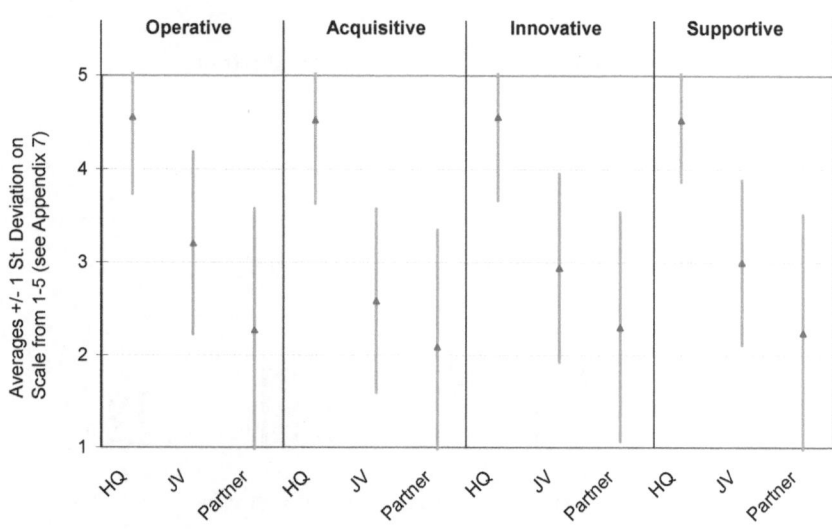

Source: Survey results.

Similarly to the results regarding technological resources, the results show that on average, the technological capabilities of the German headquarters exceed those of both the joint venture as well as the Chinese partner. Standard deviations are very high and there are several cases in which the Chinese partner's capabilities exceed those of the German headquarters (see Table 24 in Section IV-2-3).

Next, we present descriptive results relating to the Chinese partner's technological insight into the activities of the joint venture. On the one hand, it is taken into account whether the Chinese partner understands the technologies employed by the JV and/or whether the partner firm could produce the JV's products by itself (these two concepts make up the variable *PartnerInsight* discussed earlier).[103] On the other hand, we have recorded the respondents'

---

[103] As discussed in Section IV-1-2-4, *PartnerInsight* is computed by calculating the arithmetic mean of the answer to the question: "The Chinese partner understands all of the technologies employed by the JV" and "The Chinese partner could produce the JV's products himself."

opinion on whether the Chinese partner's knowledge surpasses their own in relevant technological areas.

Figure 41: Degree of Technological Insight of the JV Partner

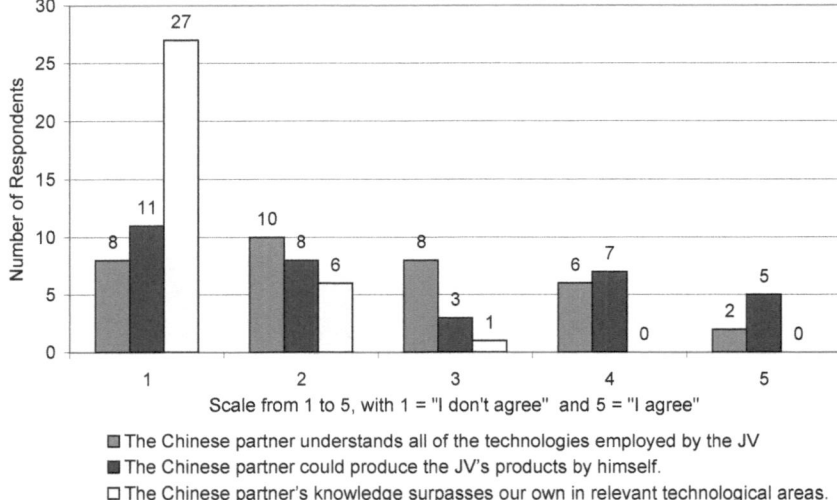

Source: Survey results.

The survey results show that there are many JVs in this sample where the Chinese partner understands the technologies employed by the joint venture and/or could produce the JV's products by himself. On the other hand, no responding manager believes that the Chinese partner's knowledge surpasses the German firm's knowledge in relevant technological areas.

One very interesting and relevant finding is that the degree of insight by the Chinese partner is not correlated to the JV's absolute or relative endowment levels. As the figure below shows for the example of *HumanwareRel*, the joint venture's relative endowment with technologically skilled human resources and the degree of insight of the Chinese partner seem to be largely independent.

Figure 42: *PartnerInsight* and *HumanwareRel*

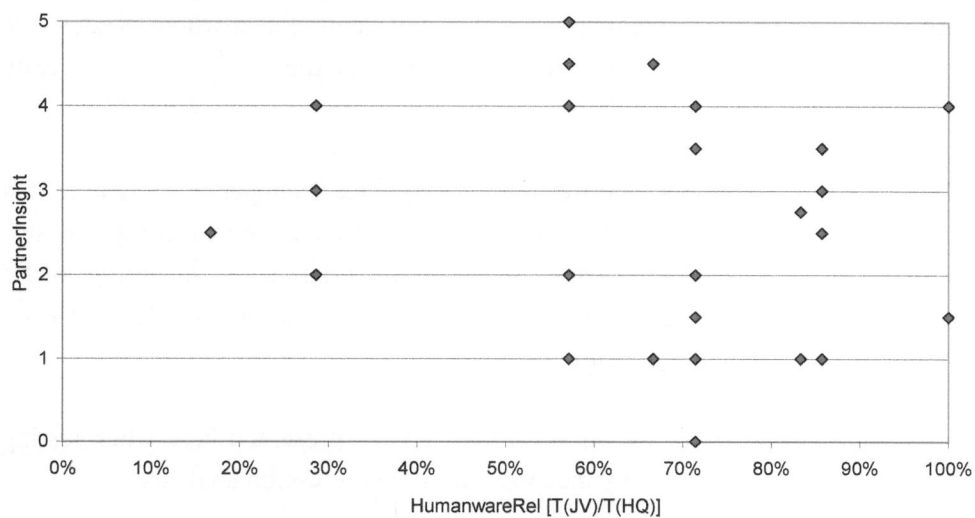

Source: Survey results.

This finding can be confirmed for all of the dimensions used in this study to measure T(JV) and T(JV)/T(HQ). The table below gives their correlation to *PartnerInsight* and the respective statistical significance levels.

Table 23: Correlations of Measures for *T(JV)* **and** *T(JV)/T(HQ)* **to** *PartnerInsight*

	Correlation	Significance (2-tailed)		Correlation	Significance (2-tailed)
TechnowareJV	0,26	0,15	TechnowareRel	0,13	0,48
InforwareJV	0,11	0,53	InforwareRel	0,13	0,47
HumanwareJV	-0,19	0,29	HumanwareRel	-0,08	0,65
OrgawareJV	-0,28	0,11	OrgawareRel	-0,21	0,23
CapabilitiesJV	-0,31	0,08	CapabilitiesRel	0,10	0,57

Source: Survey results.

As the results show, no correlation is significant, and half of the variables representing the JV's technological sophistication are even negatively correlated to *PartnerInsight*.

This is an important finding that influences the way that this study's hypotheses are tested. Joint ventures are rather interactive market entry forms and JV partners always share a certain degree of technological insight (see discussion in

Section II-3). However, when investigating the degree of technology that a German investor might share with a Chinese JV partner, it is not enough to measure the technology endowment of the joint venture as a whole. Rather one has to explicitly take into account what technology the Chinese partner really gets access to.

This is not only to be expected from previously cited management literature that recommends controlling the JV partner's degree of technological insight. It also follows from qualitative information regarding the relative responsibilities of JV partners in this survey. As Appendix 6 shows, a JV partner is often not even present at the common organisation site.

In the following section, we build on the findings presented here when testing the hypotheses of this study by means of multiple regression analysis.

2.3. Statistical Investigation of Hypotheses

After a preliminary descriptive analysis of this survey's sample data, we are now in a position to test the previously derived theoretical hypotheses. The results we present in this section offer statistical evidence relating to our hypotheses, taking into account the general factors discussed in Section III-2.

Results are generally based on multiple regression analysis. In order to implement that method, the concepts that relate to both the hypotheses as well as general factors have been defined as variables. Definitions and descriptive statistics of these variables are supplied in Appendix 4 (hypotheses) and Appendix 5 (general factors).

One limiting factor for the application of multiple regression analysis in this approach is the small number of cases. The general dataset comprises 34 observations, which prohibits the computation of comprehensive regressions including the whole set of hypotheses. We therefore test each variable relating to a certain hypothesis separately as an extension of a base model.

One Key Assumption

The following results rest on a key assumption that relates to the logic of measurement. This assumption and its consequences for the analysis shall be shortly explained before the results are presented.

The assumption made in this study is that *unless the technological sophistication of the Chinese partner – as described by the respondent – exceeds the technological sophistication of either the German firm or the joint venture, the maximum technological level of the joint venture is assumed to be an endowment by the German firm and thus a result of international technology transfer efforts.*

If the assumption above holds, one can use the technology sophistication levels of joint ventures and the values of T(JV)/T(HQ) of each joint venture as can an indirect indicator of technology commitment by German firms towards their respective joint ventures.

However, this assumption prohibits the inclusion of cases where the technological sophistication of the Chinese partner exceeds the sophistication level of the German firm worldwide or the joint venture. The table below shows the number of cases per measurement dimension for which this occurs.

Table 24: Cases in Sample for which the Sophistication Level of P exceeds HQ or JV

Partner exceeds..	Joint Venture	German firm
in Technoware.	0	0
in Inforware.	1	0
in Humanware.	5	0
in Orgaware.	7	1
in Capabilities (avg).	9	1

Source: Survey results. N =34.

For example, the absolute level of humanware of the Chinese partner exceeds the absolute level of the joint venture in five cases. This reduces the applicable data set for measuring the dimension humanware from 34 to 29.

Possible limitations due to this assumption are further elaborated on in Section V-II. At this stage, we just point out implications for reducing the final data set and the result in itself: there seem to be certain dimensions of technological sophistication in which the Chinese partner is more likely to exceed the foreign investor firms than in others.

That our assumption holds, i.e. that the maximum technological sophistication level of any joint venture under observation is the result of technology transfer by the German firm, can be verified by reading the qualitative descriptions of survey respondents regarding the role and responsibilities of JV partners (see Appendix 6). In these descriptions, there exists no case in which the endowment of a JV with sophisticated technology can be traced back to the Chinese JV partner.[104]

The next paragraphs are structured as follows. For each relevant dimension that can be used to describe the technological endowment of a joint venture or indirectly the technology transfer commitment by the German investor firm, i.e. technoware, inforware, humanware, orgaware and capabilities, we discuss all possible measurement concepts for the dependent variable and corresponding statistical results. We summarise and discuss all results in Section IV-3.

2.3.1. Findings related to Technoware

The first section discusses the resource endowment of a JV with technoware. Technoware has been defined as the tangible and palpable part of the machineries that an organisation is endowed with. It comprises physical equipment such as plants, machines, or even single tools (see Chapter II). Study participants were asked to indicate the maximum level of technoware that is present in the German firm worldwide (HQ), the joint venture (JV) and the Chinese partner on a scale from one to seven.

---

[104] Unfortunately, two descriptions are missing. However, these two cases are not critical as the technology sophistication level of the Chinese partner is rather low.

In order to shed light on the dimension Technoware, we first present statistical results of a base model that is used to predict *T(JV)*, i.e. any joint venture's *absolute* endowment with technology. We then go on to describe models for *T(JV)/T(HQ)* and *[PartnerInsight*T(JV)/T(HQ)]*. The corresponding names used are *TechnowareJV*, *TechnowareRel*, and *Shared_Technoware*.

TechnowareJV

The following table shows the base model that incorporates the best-fitting predictors of *TechnowareJV* when drawing on the general factors discussed in Section III-2. It also includes an extended model that incorporates one variable derived from the set of hypotheses.

Table 25: Regression Results for *TechnowareJV*

TechnowareJV	Base Model				+ Transparency			
	B	SE	T	Sign.[1]	B	SE	T	Sign.[1]
(cons)	0,119	1,390	0,086		-1,379	1,420	-0,971	
TechnowareHQ	0,505	0,138	3,674	***	0,466	0,127	3,660	***
TechnowareP	0,309	0,088	3,504	***	0,233	0,087	2,671	**
FirmSize	-0,020	0,200	-0,108		0,009	0,190	0,501	
GlobalIntegr	-0,333	0,182	-1,836	*	-0,214	0,174	-1,231	
Replicability	0,826	0,269	3,066	***	0,892	0,249	3,588	***
PreferenceJV	-0,679	0,372	-1,825	*	-0,840	0,348	-2,415	**
Transparency					0,422	0,176	2,393	**

[1] Results are significant at the 1% level (***), at the 5% level (**), or at the 10% level (*).

Model Statistics:

Sign. Model (F-test):	0,000		0,000	
R^2-adjusted:	0,626		0,686	
n	32		32	

Source: Survey results.

The results show that a joint venture's absolute endowment is positively related to the absolute level of technoware of the German firm worldwide (*TechnowareHQ*) and the absolute level of technoware of the Chinese partner (*TechnowareP*). This was to be expected, because the absolute level of firm assets controls for a number of firm characteristics, such as the industry-specific requirements for machinery. *FirmSize*, however, a variable that measures the

German investors' firm size as approximated by the number of employees, is not significant.[105]

Besides these three basic variables, the variables that best explain *TechnowareJV* from the set of general factors available are *GlobalIntegr*, *Replicability*, and *PreferenceJV*. The more a foreign investor is committed to global integration and the standardisation of activities – as opposed to the local adaptation of activities – the lower is the level of technoware in the JVs under observation. On the other hand, the higher the replicability of a certain technology used for production, the higher the level of technoware in a JV. Finally, a surprising result is that JVs that represent a 'free choice' of the foreign investor regarding the market entry form (*PreferenceJV* = 1) exhibit a *lower* average level of technoware than JVs that were founded due to investment regulations (*PreferenceJV* = 0).

The base model described here has overall statistical significance at the 1% level (F-test), and reaches an adjusted R^2 of almost 63%. This can be considered as rather high.

On top of this model, it is now possible to add individual variables that have been designed to test the hypotheses of this thesis.[106] As the table shows, one variable that can enhance the base model quality is the variable *Transparency*. Including *Transparency* in the regression leads to an adjusted R^2 of 68,6% and the individual variable *Transparency* is significant at the 5% level. However, including transparency into the model diminishes the significance of *GlobalIntegr*. None of the other possible variables relating to the hypotheses (see Appendix 4) are found to be significant. For the sake of brevity, these results are not shown, but a final summary table in Section IV-3 will take account of them.

---

[105] *FirmSize* is computed by dividing the number of employees of a firm by 10.000. It is neither significant when excluding *TechnowareHQ* and not even significantly correlated to *TechnowareJV* when just measuring bivariate correlations.

[106] A detailed description of these variables can be found in Appendix 4.

184

TechnowareRel

The next table refers to results concerning *TechnowareRel*, the dependent variable that represents the measure *T(JV)/T(HQ)* for the dimension Technoware. When assessing the JV's technology level, it takes the absolute level of the German headquarters into account and therefore represents a measure for the *relative* investment in a JV's technological assets. We include regression results for the base model as well as for the previously used variable Transparency. As the dependent variable *TechnowareRel* incorporates the German headquarters' sophistication level, *TechnowareHQ* has been removed as an independent variable.

Table 26: Regression Results for *TechnowareRel*

TechnowareRel	Base Model				+ Transparency			
	B	SE	T	Sign.[1]	B	SE	T	Sign.[1]
(cons)	0,547	0,178	3,079	***	0,367	0,219	1,679	
TechnowareP	0,032	0,014	2,286	**	0,023	0,015	1,526	**
FirmSize	-0,020	0,003	-0,771		-0,001	0,003	-0,444	
GlobalIntegr	-0,069	0,030	-2,308	**	-0,058	0,030	-1,898	
Replicability	0,168	0,042	3,968	***	0,178	0,042	4,205	***
PreferenceJV	-0,074	0,061	-1,212		-0,088	0,061	-1,446	
Transparency					0,043	0,032	1,368	

[1] Results are significant at the 1% level (***), at the 5% level (**), or at the 10% level (*).

Model Statistics:

Sign. Model (F-test):	0,000	0,001
R^2-adjusted:	0,475	0,492
n	32	32

Source: Survey results.

As the results above show, the relative measurement of the investment of technoware does not add much insight in this case. For the base model, the key difference is that *PreferenceJV* loses significance. For the extended model, neither the variable *Transparency* nor the variable *PreferenceJV* are significant. None of the possible variables relating to the hypotheses are found to be significant. For the sake of brevity, these results are not shown.

Shared_Technoware

Finally, the measure *Shared_Technoware* is being investigated. This is computed by multiplying *TechnowareRel* with *PartnerInsight* and thus

185

represents a measure for the relative asset proportion that is not only invested into the JV, but actually shared with the JV partner. As the results below show, the base model for *Shared_Technoware* is different as it takes account of the different logic of the dependent variable.

Table 27: Regression Results for *Shared_Technoware*

Shared_ Technoware	Base Model				+ Vs			
	B	SE	T	Sign.[1]	B	SE	T	Sign.[1]
(cons)	-1,253	0,673	-1,861	*	-1,902	0,705	-2,699	*
TechnowareP	0,157	0,088	1,788	*	0,184	0,083	2,205	*
FirmSize	-0,007	0,017	-0,409		-0,005	0,015	-0,321	
Replicability	0,556	0,219	2,541	**	0,604	0,205	2,949	***
Absorptive	0,412	0,131	3,146	***	0,298	0,129	2,304	**
Vs					0,312	0,123	2,536	**
LengthContract								
Dependency								

[1] Results are significant at the 1% level (***), at the 5% level (**), or at the 10% level (*).

Model Statistics:

Sign. Model (F-test):	0,000	0,000
R²-adjusted:	0,536	0,585
n	31	31

(table continued)

Shared_ Technoware	+ LengthContract				+ Dependency			
	B	SE	T	Sign.[1]	B	SE	T	Sign.[1]
(cons)	-0,716	0,778	-0,919		-1,621	0,729	-2,224	*
TechnowareP	0,240	0,108	2,216	**	0,138	0,088	1,571	
FirmSize	0,006	0,018	0,352		-0,002	0,017	-0,099	
Replicability	0,470	0,252	1,867	*	0,543	0,216	2,514	**
Absorptive	0,311	0,153	2,034	*	0,373	0,131	2,851	***
Vs								
LengthContract	-0,010	0,006	-1,735	*				
Dependency					0,196	0,113	1,738	*

[1] Results are significant at the 1% level (***), at the 5% level (**), or at the 10% level (*).

Model Statistics:

Sign. Model (F-test):	0,001	0,000
R²-adjusted:	0,516	0,532
n	26	30

Source: Survey results.

The base model again takes into account of the partner's absolute level of technoware (*TechnowareP*), *FirmSize* and *Replicability*. The results show that a Chinese partner's resource endowment with technoware and the replicability of a certain technology are positively related to *Shared_Technoware*, whereas *FirmSize* is again insignificant.

The base model also controls for the general absorptive capability for technology of the Chinese JV partner (*Absorptive*). It is straightforward to argue that a higher absorptive capability for technology in general leads to a high degree of insight by the partner for the certain technology described at hand (Cohen and Levinthal 1990). The combined base model is significant as a whole and reaches an R^2-adjusted of almost 54%.

On top of the base model, several variables relating to our hypotheses add to the quality of the results. *Vs*, the expected synergy effects by sharing technology drive up *Shared_Technoware*. This lends support to the hypothesis 1d. *LengthContract,* the total JV contract length in years, negatively corresponds with shared technological insight. This result contradicts hypothesis 3b that predicts a positive influence. A drawback here is that *LengthContract* has only 26 cases and thus reduces the overall quality of the regression model. Finally, *Dependency* has been found to exert a positive influence on the level of *Shared_Technoware*. As hypothesis 6b predicts, the higher the mutual dependency of the German investor firm and the Chinese partner after the JV contract under observation ends, the higher the technology sharing commitment by the German partner

The next section describes corresponding results for the dimension inforware.

2.3.2. Findings related to Inforware

This section discusses the resource endowment of a joint venture with inforware. Inforware has been defined as document-embodied technology, i.e. the accumulated knowledge about a technology that can be documented, such as documentation regarding the production process and related background information (see Chapter II). Study participants were asked to indicate the maximum level of inforware that is present in the German firm worldwide (HQ), the joint venture (JV) and the Chinese partner on a scale from one to seven.

In order to shed light on the dimension inforware, we first present statistical results of a base model that is used to predict *T(JV)*, i.e. any joint venture´s *absolute* endowment with Inforware.

We then go on to describe models for *T(JV)/T(HQ)* and [*PartnerInsight*T(JV)/T(HQ)*]. The respective variable names used are *InforwareJV*, *InforwareRel*, and *Shared_Inforware*.

InforwareJV

The following table shows the base model that incorporates the best-fitting predictors for *InforwareJV* when drawing on the general factors discussed in Section III-2. It also includes an extended model that incorporates one variable derived from the set of hypotheses, namely *DiscountRate*.

Table 28: Regression Results for *InforwareJV*

InforwareJV	Base Model				+ DiscountRate			
	B	SE	T	Sign.[1]	B	SE	T	Sign.[1]
(cons)	2,867	1,777	1,613		-2,126	5,137	-0,414	
InforwareHQ	0,571	0,247	2,310	**	0,518	0,714	0,725	
InforwareP	0,161	0,111	1,455		-0,260	0,180	-1,444	
FirmSize	-0,150	0,022	-0,665		0,037	0,021	1,700	
EquityJV	-1,644	0,710	-2,315	**	1,071	1,830	0,585	
PreferenceJV	-0,911	0,474	-1,924	*	-0,631	0,647	-0,974	
DiscountRate					0,278	0,108	2,578	**

[1] Results are significant at the 1% level (***), at the 5% level (**), or at the 10% level (*).

Model Statistics:

Sign. Model (F-test):	0,028	0,033
R²-adjusted:	0,245	0,577
n	32	15

Source: Survey results.

The results for the base model show a different logic than for *TechnowareJV*. The absolute endowment of a joint venture with inforware is significantly related to the level of inforware at the German headquarters, but not to the level of inforware of the Chinese partner. Furthermore, the level of *InforwareJV* depends significantly on the JV type. The few contractual joint ventures included in the sample (n=3) exhibit a significantly higher level of inforware than the equity JVs (n=31). This is intuitively correct as contractual JVs are mainly about passing on information that can be documented (see discussion in Chapter II). As previously found for the dimension technoware, *PreferenceJV* has a significantly negative effect on *InforwareJV*: JVs that were supposed to be

WFOEs but were implemented as JVs exhibit a significantly higher level of *InforwareJV*. The fit of the base model is not extremely good overall (R^2-adjusted = 24,5%), but acceptable.

The quality of fit increases substantially if the base model is extended by the variable *DiscountRate*, representing the discount rate applied for the JV project by the German investor firm. High discount rates significantly correspond with high levels of *InforwareJV*. This contradicts the stated hypothesis 4a. The argument underlying hypothesis 4a is that an investor with a low interest rate assigns a lot of weight to future payments relative to investors who apply a high discount factor. Such an investor would therefore be less willing to sacrifice future payoffs by defecting on the partner at the present time in order to make immediate gains. This mechanism seems to be wrong, however. An alternative explanation in line with the empirical results could be that an investor who applies a high discount rate does not value the future payoffs from the asset *InforwareJV* as much as other investors and thus invests it more deliberately even if technology diffusion might occur.

One drawback of using *DiscountRate* is that the discount rate applied to a foreign project has been treated very confidentially by survey participants, resulting in only 15 observations. This affects the overall quality of the model and the significance of all the other variables included in the base model.

InforwareRel

The next table refers to results concerning *InforwareRel*, the dependent variable that represents the measure *T(JV)/T(HQ)* for the dimension Inforware. As described for *TechnowareRel*, it takes the absolute level of the German headquarters into account and therefore represents a measure for the *relative* investment of technological assets. The next table shows regression results for the base model as well as the extension by *DiscountRate*. As the dependent variable *InforwareRel* incorporates the headquarters' sophistication level, *InforwareHQ* has been removed as an independent variable.

Table 29: Regression Results for *InforwareRel*

InforwareRel	Base Model				+ *DiscountRate*			
	B	SE	T	Sign.[1]	B	SE	T	Sign.[1]
(cons)	1,016	0,127	8,007	***	0,236	0,385	0,613	
InforwareP	0,024	0,017	1,400		-0,036	0,024	-1,516	
FirmSize	-0,003	0,003	-0,815		0,005	0,003	1,787	
EquityJV	-0,245	0,111	-2,213	**	0,144	0,243	0,590	
PreferenceJV	-0,148	0,074	-2,015	*	-0,095	0,085	-1,121	
DiscountRate					0,039	0,014	2,753	**

[1] Results are significant at the 1% level (***), at the 5% level (**), or at the 10% level (*).

Model Statistics:

Sign. Model (F-test):	0,087	0,016
R^2-adjusted:	0,142	0,601
n	32	15

Source: Survey results.

The results are very similar to the results for *InforwareJV*. However, the R^2-adjusted is considerably lower, *EquityJV* loses some significance, and *PreferenceJV* gains some significance.

Similarly, when *DiscountRate* is added, the model gains explanatory power, but at the cost of losing 17 cases. No general variable then remains significant.

Shared_Inforware

The measure *Shared_Inforware* is computed by multiplying *InforwareRel* with *PartnerInsight* and thus represents a measure for the relative proportion of inforware that can be accessed by the Chinese JV partner. The base model for *Shared_Inforware* and its extensions are shown in the following table.

Table 30: Regression Results for *Shared_Inforware*

Shared_ Inforware	Base Model				+ Vs				+ FreqInteract			
	B	SE	T	Sign.[1]	B	SE	T	Sign.[1]	B	SE	T	Sign.[1]
(cons)	1,360	0,637	2,137	**	1,497	0,689	2,172	**	0,784	0,852	0,921	
InforwareP	-0,018	0,108	-0,165		-0,050	0,097	-0,514		-0,083	0,105	-0,794	
FirmSize	0,000	0,015	0,030		0,005	0,013	0,369		0,000	0,014	-0,019	
EquityJV	-1,077	0,488	-2,206	**	-1,594	0,530	-3,009	***	-1,320	0,566	-2,334	**
PreferenceJV	-0,433	0,348	-1,242		-0,527	0,309	-1,704		-0,343	0,325	-1,054	
Absorptive	0,634	0,144	4,411	***	0,512	0,129	3,969	***	0,541	0,137	3,959	***
Vs					0,348	0,113	3,083	***				
FreqInteract									0,387	0,164	2,357	**
Links												
Dependency												

[1] Results are significant at the 1% level (***), at the 5% level (**), or at the 10% level (*).

Model Statistics:

Sign. Model (F-test):	0,000	0,000	0,000
R²-adjusted:	0,546	0,668	0,662
n	31	30	30

(table continued)

Shared_ Inforware	+ Links				+ Dependency			
	B	SE	T	Sign.[1]	B	SE	T	Sign.[1]
(cons)	1,613	0,720	2,241	**	1,048	0,781	1,342	
InforwareP	-0,094	0,105	-0,893		-0,139	0,108	-1,283	
FirmSize	-0,012	0,029	-0,399		0,009	0,014	0,640	
EquityJV	-1,502	0,549	-2,738	**	-1,373	0,556	-2,468	**
PreferenceJV	-0,561	0,350	-1,604		-0,074	0,341	-0,218	
Absorptive	0,598	0,130	4,617	***	0,571	0,132	4,324	***
Vs								
FreqInteract								
Links	0,303	0,114	2,667	**				
Dependency					0,292	0,116	2,518	**

[1] Results are significant at the 1% level (***), at the 5% level (**), or at the 10% level (*).

Model Statistics:

Sign. Model (F-test):	0,000	0,000
R²-adjusted:	0,651	0,632
n	29	30

Source: Survey results.

As the table above shows, the dependent variable *Shared_Inforware* is already explained by the base model to a high degree. The general absorptive capability of the JV partner is again an important control factor (*Absorptive*). However, there are four individual variables derived from the hypotheses that individually increase the quality of the model considerably.

All else being equal, high expected synergy gains from knowledge sharing (*Vs*) by the German investor increase *Shared_Inforware* significantly, lending support to hypothesis 1d. Secondly, a high frequency of interaction between the JV partners (*FreqInteract*) corresponds with high levels of *Shared_Inforware*.

This finding supports hypothesis 5a. Third, the presence among organisational *Links* between the JV partners apart from the JV under observation increases *Shared_Inforware*, supporting hypothesis 6a. Hypothesis 6b is also supported, stating that the expectation of future *Dependency* on the JV partner – even after the JV contract has ended – will positively influence the degree of technology shared by the German investor.

To summarise the findings related to Inforware, one can argue that while the absolute measure *InforwareJV* is well-explained by general influence factors, the measure for relative sharing of technology *Shared_Inforware* is significantly improved by predictors that relate to the cooperation dynamics of a joint venture. Therefore, the investigation of the influence factors on *Shared_Inforware* results in interesting insights regarding the hypotheses of this thesis.

2.3.3. Findings related to Humanware

This section discusses the endowment of a JV's resources with humanware. Humanware has been defined as person-embodied technology and refers to the human resources that an organisation can draw on to realise the potential of technoware, such as the skills and experiences of people engaged in the organisation (see Chapter II). Study participants were asked to indicate the maximum level of humanware that is present in the German firm worldwide (HQ), the joint venture (JV) and the Chinese partner on a scale from one to seven.

In order to shed light on the dimension Humanware, we first present statistical results of a base model that is used to predict $T(JV)$, i.e. any joint venture's *absolute* endowment with Humanware. We then go on to describe the implementation of $T(JV)/T(HQ)$ and then the implementation of $[PartnerInsight*T(JV)/T(HQ)]$. The respective variable names used are *HumanwareJV*, *HumanwareRel*, and *Shared_Humanware*.

HumanwareJV

The following table shows the base model that incorporates the best-fitting predictors of *HumanwareJV* when drawing on the general factors discussed in Section III-2. It also includes two extended models that incorporate the variables *Cac* and *AgeJV*.

Table 31: Regression Results for *HumanwareJV*

HumanwareJV	Base Model				+ Cac				+ AgeJV			
	B	SE	T	Sign.[1]	B	SE	T	Sign.[1]	B	SE	T	Sign.[1]
(cons)	0,901	3,837	0,235		1,622	3,589	0,452		-3,412	3,978	-0,858	
HumanwareHQ	-0,153	0,605	-0,252		-0,097	0,564	-0,172		0,327	0,611	0,536	
HumanwareP	0,098	0,122	0,805		0,203	0,126	1,605		0,070	0,118	0,592	
FirmSize	-0,035	0,021	-1,676		-0,032	0,019	-1,642		-0,049	0,019	-2,581	**
GovIncentives	0,596	0,192	3,103	***	0,486	0,188	2,588	**	0,445	0,180	2,480	**
OwnershipShare	0,035	0,011	3,183	***	0,032	0,010	3,081	***	0,042	0,010	4,272	***
CompPressure	0,352	0,177	1,985	*	0,329	0,165	1,989	*	0,408	0,158	2,585	**
Cac					-0,384	0,205	-1,871	*				
AgeJV									0,097	0,041	2,348	**

[1] Results are significant at the 1% level (***), at the 5% level (**), or at the 10% level (*).

Model Statistics:

Sign. Model (F-test):	0,005	0,003	0,002
R^2-adjusted:	0,512	0,578	0,664
n	23	23	21

Source: Survey results.

As the empirical results show, the optimal base model for *HumanwareJV* is again different from the optimal base model for *TechnowareJV* and *InforwareJV*, taking account of the specific properties of this dimension of technological resources.

One first property to point out is that the absolute level of *HumanwareJV* seems to be unrelated to *HumanwareHQ*. Referring to the descriptive results shown earlier, this can be explained by the fact that almost all German firms indicated the level of humanware to be maximal for the headquarters. Therefore this variable doesn't actually vary and has no explanatory power for *HumanwareJV*. Second, a variable that has been dropped is *PreferenceJV*. This variable is significant and has a negative coefficient, but becomes insignificant when *GovIncentives* is added.

The resulting base model shows that the absolute level of humanware in a JV is positively related to the extent of investment incentives offered by the Chinese government. It is also positively related to the relative JV ownership share of the German investor (*OwnershipShare*). Finally, *HumanwareJV* is positively affected by the subjectively felt degree of competitive pressure that the German investors had been exposed to when investing in China.

The base model results can be enhanced by including two factors related to the stated hypotheses. On the one hand, the level of *HumanwareJV* is negatively related to the German investor's cost of sharing the technology at hand (*Cac*). This finding supports hypothesis 1a. On the other hand, a joint venture's endowment humanware seems to be higher for old JVs (*AgeJV*). This supports hypothesis 3a. We also point out that when *AgeJV* is included in the model, the negative influence of *FirmSize* on the absolute level of humanware in a joint venture is significant on the 5% level.

The next section will discuss the corresponding results for *HumanwareRel*.

HumanwareRel

The following table refers to results concerning *HumanwareRel*, the dependent variable that represents the measure $T(JV)/T(HQ)$ for the dimension humanware. It takes the absolute level of the German headquarters into account and therefore represents a measure for the *relative* investment of technological assets. However, as described before, the absolute level of the German headquarters is almost always 7 (the maximum value), so that the incremental value of controlling for *HumanwareHQ* is relatively small. The next table shows regression results for the base model as well as the extensions by *Cac* and *AgeJV*.

Table 32: Regression Results for *HumanwareRel*

HumanwareRel	Base Model				+ Cac				+ AgeJV			
	B	SE	T	Sign.[1]	B	SE	T	Sign.[1]	B	SE	T	Sign.[1]
(cons)	0,194	0,118	1,636		0,373	0,137	2,73		0,065	0,121	0,533	
HumanwareP	0,011	0,019	0,552		0,030	0,020	1,508		0,006	0,019	0,336	
FirmSize	-0,004	0,003	-1,205		-0,003	0,003	-1,168		-0,005	0,003	-1,708	(0,108)
GovIncentives	0,079	0,030	2,613	**	0,060	0,029	2,068	*	0,065	0,030	2,159	**
OwnershipShare	0,004	0,002	2,532	**	0,004	0,002	2,444	**	0,005	0,002	3,344	***
Cac					-0,069	0,032	-2,144	**				
AgeJV									0,012	0,007	1,717	(0,107)

[1] Results are significant at the 1% level (***), at the 5% level (**), or at the 10% level (*).

Model Statistics:

Sign. Model (F-test):	0,009	0,004	0,004
R²-adjusted:	0,404	0,503	0,532
n	23	23	21

Source: Survey results.

The results for the measure *HumanwareRel* are very similar to the absolute measure *HumanwareJV*, for the reason that *HumanwareHQ* doesn't vary much. *GovIncentives* and *OwnershipShare* remain the best predictors of the relative degree of investment in humanware committed to by the German investor. The cost of sharing one's own technology (*Cac*) contributes to the estimation quality. Similarly, old JVs exhibit higher levels of *HumanwareRel* than younger JVs.

Shared_Humanware

The measure *Shared_Humanware* is computed by multiplying *HumanwareRel* with *PartnerInsight* and thus represents a measure for the relative proportion of humanware that can be accessed by the Chinese JV partner. The base model for *Shared_Humanware* and its extensions are shown below.

Table 33: Regression Results for *Shared_Humanware*

Shared_ Humanware	Base Model				+ FreqInteract				+ Dependency			
	B	SE	T	Sign.[1]	B	SE	T	Sign.[1]	B	SE	T	Sign.[1]
(cons)	0,113	0,396	0,285		0,296	0,726	0,408		0,501	0,645	0,777	
HumanwareP	-0,018	0,084	-0,216		0,168	0,102	1,637		0,139	0,106	1,311	
FirmSize	0,007	0,013	0,556		-0,001	0,018	-0,060		0,007	0,019	0,359	
GovIncentives	-0,014	0,119	-0,117		-0,099	0,165	-0,601		-0,041	0,171	-0,239	
Absorptive	0,558	0,111	5,005	***								
FreqInteract					0,373	0,192	1,947	*				
Dependency									0,29	0,15	1,933	*

[1] Results are significant at the 1% level (***), at the 5% level (**), or at the 10% level (*).

Model Statistics:

Sign. Model (F-test):	0,000	0,133	0,135
R^2-adjusted:	0,548	0,137	0,135
n	27	26	26

Source: Survey results.

As it turns out, the model quality for *Shared_Humanware* is rather low. Although the base model reaches an adjusted R^2 of over 50% and is significant, the explanatory power of the model relies on the variable *Absorptive*, which has been shown to be an important control variable in earlier models. No other significant control variable could be added to improve the model quality. Furthermore, the variables *FreqInteract* and *Dependency* can be added to the model and they each reach a significance level of 10% provided that *Absorptive* is left out. However, the two extended models are questionable as a whole (significance level of F-test is over 10%). Therefore, we summarise the findings regarding *Shared_Humanware* as very weak support for the two hypotheses relying on *FreqInteract* and *Dependency*. No significant influence could be shown in this case.

The next section will discuss the dimension orgaware.

2.3.4. Findings related to Orgaware

Orgaware has been defined as the component of technology that is embodied in institutions and refers to the supporting net of principles, practices, and arrangements that govern the effective use of the other three components. The way this study implements the dimension Orgaware, it is very much related to the *size* of an organisational entity (see Appendix 2).

The regression results that are presented in the next paragraphs reflect that fact. Because the size of a commonly-owned joint venture is not necessarily an expression of knowledge sharing, the variables derived from the hypotheses turn out to add little to predict *OrgawareJV*, *OrgawareRel* and *Shared_Orgaware*. In fact, although a base model for the measure *Shared_Orgaware* is shown, we suggest that that model should not be taken into account because the logic behind it is not straightforward.

OrgawareJV

The institutional sophistication and size of the joint venture organisation (*OrgawareJV*) can be predicted by the variables included in the model shown below.

Table 34: Regression Results for *OrgawareJV*

OrgawareJV	Base Model				+ TimeHorizon			
	B	SE	T	Sign.[1]	B	SE	T	Sign.[1]
(cons)	5,210	1,847	2,821	**	5,755	1,768	3,255	***
OrgawareHQ	0,275	0,212	1,296		0,224	0,204	1,096	
OrgawareP	0,248	0,169	1,464		0,431	0,185	2,330	**
FirmSize	0,102	0,052	1,980	*	-0,159	0,106	-1,500	
PreferenceJV	-1,332	0,531	-2,506	**	-1,338	0,492	-2,717	**
RelResp	-2,149	0,870	-2,47	**	-2,938	0,857	-3,428	***
RelResp^2	0,299	0,168	1,776	*	0,407	0,157	2,584	**
TimeHorizon					0,143	0,074	1,921	*

[1] Results are significant at the 1% level (***), at the 5% level (**), or at the 10% level (*).

Model Statistics:

Sign. Model (F-test):	0,000	0,001
R^2-adjusted:	0,628	0,778
n	26	19

Source: Survey results.

The base model results show that *FirmSize* has a positive influence on *OrgawareJV*.[107] Furthermore, *PreferenceJV* is significantly negative, indicating that organisations that were founded as JVs as a result of free choice are smaller

---

[107] The measures *OrgawareJV* and *FirmSize* are also positively correlated (+0,28), but not to the extent that collinearity prohibits the inclusion of both. N equals 26, because there are seven cases in which the Chinese firm's level in orgaware exceeds either *OrgawareHQ or OrgawareJV*.

than organisations that were founded as JVs due to investment regulations. Finally, the variables *RelResp* and *RelResp^2* are significant.[108] The negative sign for *RelResp* indicates the following relationship: all else being equal, the higher the responsibility of the Chinese partner for the functional activities 'research', 'development', 'sourcing' and 'production', the lower is *OrgawareJV*. In other words, in cases where the German firm does these activities itself, the Chinese JV is usually bigger. *RelResp^2* is an additional squared term controlling for a non-linear relationship. The results show that *RelResp^2* has a significantly positive relationship to *OrgawareJV*, but the relationship between *RelResp* and *OrgawareJV* is not further discussed here for the sake of brevity.

Regarding the evaluation of hypotheses, the base model can only be extended by the variable *TimeHorizon*, which represents a test of hypothesis 4b. This variable is significant at the 10% level without having a negative impact on the other results, even though the sample size decreases to 19. One can therefore establish the relationship that firms who apply a long time horizon for the investment in a JV establish a larger organisation (which seems quite logical).

The next section considers findings related to *OrgawareRel*.

OrgawareRel

Regarding *OrgawareRel*, the application of the concept *T(JV)/T(HQ)* for the dimension Orgaware, the results are very similar to the model for the absolute level of Orgaware. However, the extra variable *TimeHorizon* is not significant. These results are shown below.

---

[108]   *RelResp* summarises to what extent the German and the Chinese partners divide responsibility for the functional activities research, development, sourcing and production. Relative responsibility for each of these functional activities has been measured on a scale from 1 to 5, where 1 indicates 'only German firm', 3 indicates 'both' and 5 indicates 'only Chinese partner'. *RelResp^2* is the squared term of *ResResp*, accounting for a nonlinear relationship between *RelResp* and *OrgawareJV*.

Table 35: Regression Results for *OrgawareRel*

OrgawareRel	Base Model				+ TimeHorizon			
	B	SE	T	Sign.[1]	B	SE	T	Sign.[1]
(cons)	1,064	0,188	5,657	***	1,089	0,196	5,549	***
OrgawareP	0,029	0,026	1,102		0,060	0,036	1,676	
FirmSize	0,012	0,008	1,525		-0,031	0,019	-1,600	
PreferenceJV	-0,193	0,087	-2,203	**	-0,200	0,097	-2,057	**
RelResp	-0,340	0,142	-2,402	**	-0,413	0,160	-2,586	**
RelResp^2	0,056	0,028	2,035	*	0,065	0,031	2,12	*
TimeHorizon					0,016	0,014	1,096	

[1] Results are significant at the 1% level (***), at the 5% level (**), or at the 10% level (*).

Model Statistics:

Sign. Model (F-test):	0,012	0,028	
R^2-adjusted:	0,372	0,465	
n	26	19	

Source: Survey results.

One can summarise that the JV organisation size in China relative to the organisation size of the German firm is bigger for 'enforced JVs'. Also, the higher the responsibility of the German firm for the tasks 'research', 'development', 'sourcing' and 'production', the larger the relative investment in *OrgawareJV*. On the other hand, the *FirmSize* of the German firm has no influence on this relative degree of investment. *TimeHorizon* loses its significance in this model.

Shared_Orgaware

The final measure to be applied for the dimension orgaware is *Shared_Orgaware*. In the case of Orgaware, the interpretation of the measure *PartnerInsight*T(JV)/T(HQ)* is not very intuitive, which is why the results for any regression should be handled with care. We compute the results for a regression analysis for the sake of completeness.

As the results below show, a well-fitting regression takes into account the variables *FirmSize* and *RelResp^2*. Also, the higher the *CostLevels*, i.e. the investment incentive due to low input costs such as low-and high-skilled labour, raw materials and infrastructure, the higher the *Shared_Orgaware*. One interesting result is that *FirmSize* has a positive influence on *Shared_Orgaware*:

large firms are likely to invest in relatively large JVs in which the local partner also has a lot of technical insight.

An extension of the model by *Dependency* is possible, but only at the expense of neglecting the variables *CostLevels*, *RelResp* and *RelResp^2*. This is partly due to the small sample size.

Table 36: Regression Results for *Shared_Orgaware*

Shared_ Orgaware	Base Model				+ Dependency			
	B	SE	T	Sign.[1]	B	SE	T	Sign.[1]
(cons)	0,261	0,830	0,314		0,276	0,425	0,65	
OrgawareP	0,179	0,087	2,058		0,164	0,113	1,452	
FirmSize	0,107	0,023	4,567	***	0,076	0,029	2,661	***
CostLevels	0,328	0,153	2,144	**				
RelResp	-0,754	0,480	-1,571					
RelResp^2	0,191	0,096	1,977	*				
Dependency					0,22	0,108	2,043	*

[1] Results are significant at the 1% level (***), at the 5% level (**), or at the 10% level (*).

Model Statistics:

Sign. Model (F-test):	0,001	0,013
R²-adjusted:	0,568	0,318
n	25	24

Source: Survey results.

Overall, however, the extended model does not add improvement to the model estimation, reducing R^2-adjusted from 56,8% to 31,8%. Therefore, one can conclude that based on the measurement *Shared_Orgaware*, none of the hypotheses could be verified.

The next section will present results related to the measurement concept 'technological capabilities'.

2.3.5. Findings related to Capabilities

This section presents statistical results that are based on using *CapabilitiesJV*, *CapabiliesRel* and *Shared_Capabilities* as dependent variables. Technological capabilities have been defined in Section II-1-2, and the survey has adopted a description based on the categories 'operative capabilities', 'acquisitive capabilities', 'innovative capabilities', and 'supportive capabilities' as proposed by Ramanathan (1994).[109] Because these four types of categories exhibit very high levels of correlation (see Section IV-1-2), they are combined into a single scale. The resulting three measurement concepts for capabilities are *CapabilitiesJV*, *CapabilitiesRel* and *Shared_Capabilities*.

CapabilitiesJV

Starting out with measuring the absolute level of technological capabilities to be found in a joint venture, we present the following base model and possible extensions.

Table 37: Regression Results for *CapabilitiesJV*

CapabilitiesJV	Base Model				+ Trust				+ Transparency			
	B	SE	T	Sign.[1]	B	SE	T	Sign.[1]	B	SE	T	Sign.[1]
(cons)	0,575	1,060	0,543		-0,531	1,179	-0,45		0,427	0,999	0,427	
CapabilitiesHQ	0,845	0,215	3,923	***	0,933	0,211	4,421	***	0,688	0,218	3,149	***
CapabilitiesP	0,062	0,115	0,537		-0,012	0,119	-0,100		-0,039	0,120	-0,320	
FirmSize	-0,006	0,012	-0,503		-0,005	0,012	-0,443		0,004	0,013	0,343	
GlobalIntegr	-0,340	0,110	-3,076	***	-0,313	0,108	-2,907	***	-0,275	0,109	-2,521	**
Trust					0,202	0,112	1,811	*				
Transparency									0,238	0,125	1,907	*

[1] Results are significant at the 1% level (***), at the 5% level (**), or at the 10% level (*).

Model Statistics:

Sign. Model (F-test):	0,004	0,004	0,002
R²-adjusted:	0,431	0,484	0,498
n	25	24	25

Source: Survey results.

As the table above shows, the preferred model for predicting the absolute level of technological capabilities in a joint venture finds a significant relationship between the absolute level of capabilities of the German headquarters and *CapabilitiesJV*. The level of capabilities by the Chinese partner does not seem to

---

[109] Survey participants were asked to indicate the degree of sophistication for each of these four components. This was measured on a scale from 1 to 5, where 1 represented "weak" and 5 represented "strong". Survey participants on behalf of the German firms indicated their assessment for the German firm worldwide (HQ), the joint venture (JV), and the Chinese partner (P).

matter.[110] Another result is that firms committed to the worldwide integration and standardisation of activities tend to be invested in JVs with low technological capabilities, all else being equal.

The basic model can be extended by two variables related to the hypotheses of this study, namely *Trust* and *Transparency*. According to the regression results, in cases where the German partner describes the level of mutual trust as high, joint ventures tend to be well-equipped with technological capabilities. Also in cases where the German partner assesses the behavioural transparency within the JV as high, the sophistication of JVs regarding technological capabilities is high. These results are similar for the measure *CapabilitiesRel*, as the next section will show.

CapabilitiesRel

The results for the measurement concept *CapabilitiesRel* have been summarised in the table below. The major difference to the previous results is that *Trust* is not a significant variable in this context. On the other hand, the significance of *Transparency* rises slightly, resulting in a 5% significance level.

Table 38: Regression Results for *CapabilitiesRel*

CapabilitiesRel	Base Model				+ Trust				+ Transparency			
	B	SE	T	Sign.[1]	B	SE	T	Sign.[1]	B	SE	T	Sign.[1]
(cons)	0,978	0,104	9,45	***	0,856	0,133	6,454	***	0,801	0,126	6,341	***
CapabilitiesP	0,013	0,024	0,558		0,000	0,026	0,018		-0,008	0,024	-0,320	
FirmSize	-0,001	0,002	-0,257		0,000	0,002	-0,127		0,001	0,002	0,501	
GlobalIntegr	-0,076	0,023	-3,352	***	-0,071	0,023	-3,074	***	-0,065	0,022	-2,990	***
Trust					0,034	0,023	1,46					
Transparency									0,051	0,024	2,155	**

[1] Results are significant at the 1% level (***), at the 5% level (**), or at the 10% level (*).

Model Statistics:

Sign. Model (F-test):	0,022	0,034	0,008
R^2-adjusted:	0,272	0,281	0,397
n	25	24	25

Source: Survey results.

---

[110] This conclusion is only valid for the given dataset, which excludes nine cases where the technological capabilities of the Chinese partner exceed those of the joint venture or the German partner firm.

As the interpretation of these results should be straightforward, we proceed by presenting results relating to *Shared_Capabilities*.

Shared_Capabilities

Finally, the measure *Shared_Capabilities* is investigated. This is computed by multiplying *CapabilitiesRel* with *PartnerInsight* and thus represents a measure for the degree of technological capabilities actually shared between the German firm and the Chinese JV partner. As the results below show, the base model for *Shared_Capabilities* takes account of the investment incentive due to low cost levels (*CostLevels*) as well as the general technological absorptive capacity of the Chinese JV partner (*Absorptive*).

Table 39: Regression Results for *Shared_Capabilities*

CapabilitiesRel	Base Model				+ Links				+ FreqInteract			
	B	SE	T	Sign.[1]	B	SE	T	Sign.[1]	B	SE	T	Sign.[1]
(cons)	-0,451	0,477	-0,945		-0,642	0,448	-1,433		-1,295	0,536	-2,415	**
CapabilitiesP	0,257	0,158	1,627		0,120	0,154	0,778		0,194	0,142	1,361	
FirmSize	0,013	0,019	0,652		0,025	0,018	1,404		0,016	0,017	0,940	
CostLevels	0,266	0,132	2,013	*	0,260	0,119	2,187	**	0,338	0,118	2,859	**
Absorptive	0,243	0,111	2,195	**	0,266	0,100	2,655	**	0,170	0,100	1,693	(0,11)
Links					0,192	0,079	2,443	**				
FreqInteract									0,295	0,109	2,71	**

[1] Results are significant at the 1% level (***), at the 5% level (**), or at the 10% level (*).

Model Statistics:

Sign. Model (F-test):	0,000	0,000	0,000
R^2-adjusted:	0,610	0,658	0,679
n	23	22	22

Source: Survey results.

The application of the measurement concept *Shared_Capabilities* supports two hypotheses, namely hypothesis 5a and hypothesis 6a: if the German firm maintains multiple links to the Chinese partner (*Links*), the level of shared technological insight is above average. Similarly, partnerships with a high frequency of interaction (*FreqInteract*) exhibit especially high levels of shared technological insight.[111]

---

[111] The direction of reasoning could be argued to go both ways in this case. The hypothesis is that a high frequency of interaction reduces the incentive to cheat on the partner and thus encourages cooperation and the sharing of technological capabilities.

The results for the measurement dimension 'technological capabilities' together provide some interesting insights. The explanatory power of all regressions is rather high, despite the fact that the applicable data set is small (n <= 25). Furthermore, the different measurement concepts reveal information regarding several hypotheses, namely the hypotheses regarding the influence of trust, transparency, the existence of multiple links and the frequency of interaction within the JV partnership. All of the factors mentioned point to the predicted direction, thus lending support to their respective hypotheses.

3. Summary and Discussion of Results

This section summarises and discusses the results of this study. First, we will give an overview of the most relevant results achieved and then move on to discuss them in relation to existing literature.

3.1. Summary of Results and Evaluation of Hypotheses

The qualitative and empirical results of this study offer interesting insights regarding the initially proposed research questions. We first present the general findings of this study and then systematically evaluate our hypotheses.

First, insights relate to *general characteristics* of German-Chinese international joint ventures, particularly the *cooperative setting*. The joint ventures under observation fulfil most functional activities of the value chain, with sales, production and purchasing being the most frequently observed functions. Research and development are functions less frequently observed. If a commonly owned joint venture has responsibilities for technically challenging functional activities, such as research or development, the German partner usually coordinates them. Furthermore, 22 out of 34 joint ventures were reported to be the result of 'free choice' regarding market entry, whereas in 11 cases the

---

Of course, one might also argue that the technological competence of the Chinese JV partner is a result of the frequent interaction with the German investor firm. This is further discussed in Section V-1-3.

EMPIRICAL STUDY

German investor would have preferred to use a wholly foreign-owned enterprise (WFOE).

Second, the *products* produced by the JVs in China are overwhelmingly products that have reached maturity or saturation stage of the product life cycle and are therefore not very innovative when compared to other activities of the German investor firm. However, the products are rather complex and the joint ventures produce a sizeable share of a product locally – both with respect to the physical weight and the value of a product .

Third, the analysis of a joint venture's *technological endowment* leads to interesting conclusions. Recording the sophistication levels of the German firm worldwide, the joint venture and the Chinese partner has allowed us to compute 'technology profiles' of the joint ventures. On average, the German headquarters are endowed with superior technological assets and capabilities and are the main contributor of technology to a joint venture, while the Chinese partner firm exhibits a lower technological sophistication than the joint venture. A qualitative analysis of each individual joint venture under observation underpins the argument that technologically sophisticated assets and capabilities are contributed by the German partner firm, not the Chinese partner firm.

Fourth, the technical sophistication of a joint venture does not correspond with the degree of technological insight obtained by the Chinese JV partner. All measures for sophistication were found to be uncorrelated to the variable *PartnerInsight*.

Fifth, a mixture of incentives is responsible for a German firm's commitment to invest technological resources and capabilities in a joint venture in China. The application of multiple regression analysis for the prediction of technology sophistication levels produces the following statistical results:

- The absolute size of the German investor firm is neither indicative of the absolute nor the relative level of technological sophistication. We only find weak results for the prediction of *HumanwareJV* and *OrgawareJV*.
- Firms who are committed to the *global integration* and standardisation of worldwide activities exhibit relatively low levels of technological

205

investment in their Chinese joint ventures, as could be expected following the framework by Bartlett and Ghoshal (1998). The variable *GlobalIntegr* is found to be significant for predicting the level of *TechnowareJV*, *TechnowareRel*, *CapabilitiesJV* and *CapabilitiesRel*.

- The *replicability* of a technology, representing a measure for the feasibility of technology transfer, is an important indicator. We find evidence that the higher the replicability of a certain technology, the higher the share that ends up in the joint venture and is also understood by the JV partner. This relationship can be statistically established for all dependent variables relating to technoware (*TechnowareJV*, *TechnowareRel* and *Shared_Technoware*).

- The *absorptive capability* of a Chinese partner is a strong control factor to be taken into account when predicting the level of technology eventually understood and employed by the Chinese partner. *Absorptive* takes account of the technological sophistication of the Chinese partner irrespective of the joint venture at hand, but one could also argue that a Chinese partner with a strong technological basis is more skilled in acquiring new technology (Cohen and Levinthal 1990). Statistical results support the significance of absorptive capability for most dependent variables expressing shared technological insight (*Shared_Technoware*, *Shared_Inforware*, *Shared_Humanware* and *Shared_Capabilities*).

- *Investment incentives* by the Chinese government are positively associated with current technological sophistication level of joint ventures, particularly with the sophistication level of humanware. The variable *GovIncentives* serves as a positive indicator for all three dependent variables *HumanwareJV*, *HumanwareRel* and *Shared_Humanware*.

- Joint ventures that are controlled by the German partner firm – by maintaining a high *ownership share* or by maintaining the *operational responsibility* of organisational functions, exhibit comparatively high technological sophistication levels. The variable *OwnershipShare* is statistically significant in models predicting *HumanwareJV* and *HumanwareRel*, while the variables *RelResp* and *RelResp^2* are significant for predicting *OrgawareJV* and *OrgawareRel* (and are slightly indicative of the level of *Shared_Orgaware*).

- Joint ventures that the German investor would have preferred to be WFOEs exhibit significantly higher levels of *TechnowareJV*,

TechnowareRel, InforwareJV, InforwareRel, Shared_Inforware, OrgawareJV and *OrgawareRel* than the 'free-choice-JVs'. As we have previously described, this finding is counter-intuitive from the 'cooperation' point of view, which predicts that the German investor is less willing to invest technologically sophisticated assets and capabilities in 'enforced' joint ventures, all else being equal. A possible explanation for the opposite finding is discussed in the next section.

Next, all initial hypotheses are evaluated in light of the survey results. The following table summarises the statistical evidence that supports – or contradicts – each hypothesis.[112]

---

[112] The following table provides statistical results for each of the three measurement concepts – *T(JV)*, *[T(JV)/T(HQ)]* and *Shared_Technology* – for each of the five dimensions of technological sophistication, namely technoware, inforware, humanware, orgaware and capabilities.

Table 40: Overview of Statistical Results for Evaluation of Hypotheses

Hypotheses and relating Variables		Results for: T(JV) / [T(JV)/T(HQ)] / Shared_Technology				
(Predicted direction of effect in parentheses)		Technoware	Inforware	Humanware	Orgaware	Capabilities
Hypothesis 1a - d						
1a) Costs of own cooperation	C_{AC} (-)	n/n/n	n/n/n	-*/-**/n	n/n/n	n/n/n
1b) Costs of own defection	C_{AD} (+)	n/n/n	n/n/n	n/n/n	n/n/n	n/n/n
1c) Costs of being defected upon	C_{BD} (-)	n/n/n	n/n/n	n/n/n	n/n/n	n/n/n
1d) Synergy	V_S (+)	n/n/+**	n/n/+***	n/n/n	n/n/n	n/n/n
Hypothesis 2						
Trust	Trust (+)	n/n/n	n/n/n	n/n/n	n/n/n	+*/(+)/n
Hypothesis 3a & b						
3a) Age of JV	AgeJV (+)	n/n/n	n/n/n	+**/(+)/n	n/n/n	n/n/n
3b) Duration of Contract	LengthContract (+)	n/n/-*	n/n/n	n/n/n	n/n/n	n/n/n
Hypothesis 4a & b						
4a) Discount Rate	DiscountRate (-)	n/n/n	+**/+**/n	n/n/n	n/n/n	n/n/n
4b) Time Horizon	TimeHorizon (+)	n/n/n	n/n/n	n/n/n	+*/(+)/n	n/n/n
Hypothesis 5a & b						
5a) Frequency of Interaction	FreqInteract (+)	n/n/n	n/n/+**	n/n/(+)	n/n/n	n/n/+**
5b) Behavioural Transparency	Transparency (+)	+**/n/n	n/n/n	n/n/n	n/n/n	+*/+**/
Hypothesis 6a & b						
6a) Multiple Links	Links (+)	n/n/n	n/n/+**	n/n/n	n/n/n	n/n/+**
6b) Dependency	Dependency (+)	n/n/+*	n/n/+**	n/n/(+)	n/n/+*	n/n/n

*** significant at the 1% level   * significant at the 10% level   n = no significant result
** significant at the 5% level   ( ) weak support, see description in text

Source: Survey results.

As the table above shows, the application of three measurement concepts for each of five 'technology dimensions' has provided the following evidence:

- **Hypotheses 1 a-d:**
  Two variables individually relating to the payment structure of knowledge sharing for the German investor firm seem to matter. High costs of knowledge sharing (C_{AC}) restrict the absolute and relative 'amount' of humanware present in a joint venture. This finding lends weak support to hypothesis 1a. The expected synergy gains from knowledge sharing (V_S) are positively associated with *Shared_Technoware* and *Shared_Inforware*. This supports hypothesis 1d. Evidence for or against hypothesis 1b (effect of C_{AD}) and 1c (effect of C_{BD}) could not be found.

- **Hypothesis 2:**
  The level of mutual trust within the partnership proved to be positively

related to the absolute level of technological capabilities exhibited by a joint venture (*CapabilitiesJV*). Regarding the other dimensions, trust has not turned out to be a significant variable. One could therefore say that there exists weak evidence for a link between mutual trust and the technological capabilities of a JV.

- **Hypotheses 3a and 3b:**

Evidence regarding hypotheses 3a and 3b is mixed. A positive relationship between the age of a JV and the absolute sophistication of a JVs endowment with humanware is established. This supports hypothesis 3a. However, the data contradicts hypothesis 3b: the length of a JV contract is *negatively* associated with *Shared_Technoware*, contrary to expectations. An interpretation of this result follows in the next section.

- **Hypotheses 4a and 4b:**

Hypotheses 4a and 4b can neither be rejected nor supported by the data. On the one hand, the corresponding information proved to be highly confidential or not known by the survey respondents, resulting in a restricted data set. On the other hand, the evidence is mixed. The relationship between an investor's applied discount rate and the technological sophistication is positively related to the absolute and relative level of inforware. This is opposite to the predicted relationship.

Hypothesis 4b is weakly supported, because a long time horizon corresponds with high absolute levels of orgaware in a joint venture. However, we have argued that orgaware is not the most relevant dimension in this study.

- **Hypotheses 5a and 5b:**

According to the stated hypotheses, a high frequency of interaction and high behavioural transparency contribute to fast detection of uncooperative behaviour within a relationship, which safeguards cooperative behaviour among firms. The empirical findings support this predicted effect. The variable *FreqInteract* has a positive and significant effect when included in models predicting the measurement concepts *Shared_Inforware*, *Shared_Humanware*, and *Shared_Capabilities*. The variable *Transparency* has a positive relationship to the absolute level of technoware in a JV as well as to *CapabilitiesJV* and *CapabilitiesRel*. No evidence contradicting the hypotheses could be found, so that one can accept the relationship as given. However, the problem here is the direction of causality, because both the frequency of interaction as well as the transparency of a relationship could not only be interpreted as a requirement for cooperation, but also argued to be a characteristic of close cooperation that involves intensive knowledge sharing (see discussion on causality in Chapter V).

- **Hypotheses 6a and 6b:**

  Finally, hypotheses 6a and 6b relate to the enforceability of cooperation due to multiple links or mutual dependency after the current JV contract expires. The variables related to these hypotheses exhibit a significant coefficient in the predicted direction in many models. Multiple links seem to correspond to high levels of *Shared_Inforware* and *Shared_Capabilities*. Enduring mutual dependency corresponds with high levels of *Shared_Technoware*, *Shared_Inforware*, *Shared_ Humanware* and *Shared_Orgaware*, i.e. all measures for the relative investment of technological resources that take account of the actual degree of insight of the Chinese partner.[113] For this reason, we broadly accept the mechanism proposed by hypotheses 6a and 6b. A reverse effect from a high technological sophistication of the partner on the number of links and the degree of dependency is possible, but less likely than the proposed effect.

In the next section, we discuss our results in light of the previously presented literature on the topic of international market entry, technology transfer, and international alliances.

---

[113] The evidence regarding the link between *Dependency* and *Shared_Humanware* is weak because although the coefficient of *Dependency* is significant, the F-test for the whole model results in a significance level of > 13%.

3.2. Discussion of Results

The results of this thesis represent new and useful insights for the evaluation of the two research questions initially presented:

What factors influence the sophistication of the technological endowment that an international joint venture in China receives from its German parent? and *In what way do strategic considerations regarding inter-firm cooperation and knowledge sharing influence the foreign investor's technology transfer behaviour?*

Regarding the first research question, this study provides differentiated empirical results for a variety of factors that correspond with the sophistication of an IJVs technological endowment. The factors covered shall not be repeated here, but we point out that they include characteristics of (1) the foreign investor firm, (2) the technology at hand, (3) the investment environment (most notably the strength of government incentives) and (4) characteristics of the Chinese partner firm.

Regarding the second research question, some specifically derived hypotheses that relate to the inter-firm cooperation of JV partners have been tested. All but two coefficients that are found to be significant point in the predicted direction, so that the general thought model described in Chapter III that was applied to derive our hypotheses proves to be quite relevant. The expectations formed from the analysis of theoretic models on knowledge sharing within international joint ventures have therefore be partly confirmed. We refer the reader to the previous section for an elaborate description of survey results.

There were three cases in which the expectations formed by hypotheses were contradicted by the survey results, namely for the variables *LengthContract*, *DiscountRate* and *PreferenceJV*. We discuss these variables and offer possible explanations for the contradictions below:

- *LengthContract,* the total JV contract length in years, was expected to be positively associated with technology sharing. The logic behind this

211

expectation was that the total length of a JV contract can serve as an indicator for the intended length of interaction and therefore the extent to which both partners are ready to commit resources to the common venture. However, the only statistically significant result supports a *negative* relationship (for *Shared_Technoware*). As an explanation, one could argue that a German investor's readiness to share insights with the Chinese partner decreases with the total length of the JV contract. Even though the absolute technology sophistication level of the joint venture can be high, the degree of shared technology might be low.

- Second, the *discount rate* applied to investments, representing the costs of capital and the relative weight of future payment streams, was expected to correspond negatively to the degree of technological investment.[114] Empirical tests provide contrary evidence for the relationship between *DiscountRate*, *InforwareJV* and *InforwareRel*. One alternative explanation could be that an investor who applies a high discount rate does not value the future payoffs from the asset *inforware* as much as other investors and thus invests it more deliberately even if technology diffusion might occur.

- Third, one other interesting result contradicting our expectations is that 'enforced JVs' – those joint ventures that the German investor would have preferred to form as WFOEs – exhibit *higher* absolute and relative levels of technological sophistication (in particular *TechnowareJV*, *TechnowareRel*, *InforwareJV*, *InforwareRel*, *Shared_Inforware*, *OrgawareJV* and *OrgawareRel*) than 'free choice' joint ventures. An explanation for this result is offered by the qualitative descriptions of joint ventures provided in Appendix 6: The German investor firms that wanted to use a WFOE are mostly able to *replicate* a WFOE by using a partner that is not involved in the JV's operations, e.g. a financial investor or a regional government body. The high exhibited technology sophistication level would then represent the level for the initially intended WFOE.

---

[114] According to our reasoning in Chapter III, the lower the discount rate (and thus the lower the costs of capital and the higher the weight that a firm assigns to future payoffs), the more likely a firm will engage in cooperative behaviour. We therefore suspected that JVs that are evaluated with low discount rates should exhibit more commitment by the German firm regarding its investment in sophisticated technology than those being evaluated by a high discount rate.

The results of our study also contribute insights to several strands of research previously described in Chapters II and III.

First, this study lends empirical support for the application of game theoretical arguments to the knowledge sharing between alliance partners. Previous work on static and dynamic games that has been applied to alliances (Parkhe 1993, Bruck 1996, Loebecke et al. 1998) is thus provided with another relevant piece of research. This work is closely related to Parkhe (1993) and hopefully adds insights to the matters it addresses.

Second, the finding that German investors who trust their Chinese partners endow JVs with relatively strong technological capabilities relates to literature on trust and relational capital between alliance partners. It lends empirical support to the discussion on the relevance of trust for the co-operation between organisations or organisation members (Granovetter 1985, Barney 1990, Bromiley and Cummings 1991, Ring and Van de Ven 1992, Hill 1990, Parkhe 1993, Curall and Inkpen 2002, Kale 2002).

Third, the results from this thesis can be transformed into recommendations for management practitioners and thus adds to the body of management literature on the topic (e.g. Holtbrügge and Puck 2005, von Keller et al. 2005, Kasperk et al. 2006). For example, German investor firms generally safeguard their technological assets when investing in the technological sophistication of a joint venture: they not only adjust to general influence factors regarding the market, the technology or the specific requirements of the project. They also incorporate considerations of the co-operative setting – for example the costs and benefits of knowledge sharing and the ability to react to cooperative or uncooperative behaviour. This ability for reciprocal behaviour is dependent on high behavioural transparency and future mutual dependence.

Finally, the work presented here can be used as a basis for the critical evaluation of the past application of transaction cost studies applied to the choice of a market entry form. Studies such as Weiss (1996) and Fu (2005) who investigate the choice of a market entry form simplify the occurrence 'joint venture' as a market entry form.[115]

We argue that these studies downplay that the variability of an IJV's technological sophistication is extremely high – technological sophistication should not be summarised by a dummy variable for 'R&D'. On the other hand, one should not neglect that IJVs might actually work like WFOEs, because the local partner is not involved in local operations, let alone technologically challenging tasks.

---

[115]   Typically, transaction cost studies investigating the choice of a market entry form use a dependent variable in three states, namely "hierarchical" (e.g. WFOE), "hybrid" (e.g. joint venture) and "market exchange" (e.g. Licensing). Weiss (1996) and Fu (2995) argue that a technology component (usually taking into account a dummy variable that is "1" if R&D is carried out) makes international investors choose a hierarchical market entry form due to high transaction costs.

Chapter V: Conclusion

The qualitative and empirical results of this study have successfully established a link between the co-operative setting of a German-Chinese joint venture and the sophistication of a joint venture's endowment with technological resources and capabilities.

This chapter provides our conclusion and suggestions for further research. However, the findings of this study are based on a variety of assumptions and subject to certain limitations, which will be discussed first.

1. Research Limitations

The empirical study that was carried out as part of this thesis suffers from several limitations as far as the author is aware. These limitations are a small sample size and related restrictions to statistical testing, an indirect logic applied to infer the degree of technology transferred and the ambiguity of causal chains. These three aspects are elaborated on in the following sections.

1.1. Sample Characteristics

A first drawback of this study relates to characteristics of the sample. The sample size is small and not based on random sampling, but the result of feasibility. On the one hand, it proved to be a challenge for the author to obtain the necessary information regarding German-Chinese joint ventures during the questionnaire survey.[116] On the other hand, there exists no publicly available and reliable list of German joint ventures in China. Therefore the size of the 'true population' of German-Chinese joint ventures could only be estimated.

---

[116]   There are two main reasons for this difficulty. First, the information sought after was very comprehensive and required a respondent to have strategic insight into the international operations of the respondent firm. This means that only a few people were *capable* of answering the questionnaire. Second, the information was often classified as confidential and was thus not provided.

© Springer Fachmedien Wiesbaden GmbH, part of Springer Nature 2008
M. Hoeck, *Cooperation and Technological Endowment in International Joint Ventures: German Firms in China*, Edition KWV, https://doi.org/10.1007/978-3-658-24355-5_5

The small sample size negatively affects our statistical investigation. In particular, a comprehensive model including all hypotheses could not be tested. Instead, the theoretically derived hypotheses were tested one at a time as singular additions to a base model. This is a viable option for small samples and has previously been applied by other research in the same field (e.g. Dougherty 1997).

Another principal drawback of the sample – that was already referred to when outlining the study design – is the fact that the study results are only based on answers from managers of the foreign investor firms. A more comprehensive study design would have included information sources from both the foreign and the local partner to IJVs in China. However, as was stated before, this was not feasible for a variety of reasons and the design of the empirical investigation had to cope with this limitation.

1.2. Inference of Technology Transfer Commitment

The second potential weakness of this study concerns the logical link between the hypotheses and the empirical investigation. The theoretically derived hypotheses make predictions about the cooperative commitment regarding technology transfer to be shown by a foreign investor. As described in Section IV-2-2, the testing of these predictions from observable data as implemented by this study rests on a key assumption. This assumption is that unless the technological sophistication of the Chinese partner – as described by the respondent – exceeds the technological sophistication of either the German firm or the joint venture, the maximum technological level of the joint venture is assumed to be an endowment by the German firm and thus a result of international technology transfer efforts.

This assumption is a necessary logical link to infer a preceding technology transfer commitment by a German investor firm from the technology endowment's sophistication of a certain joint venture under observation. The (relative) technology transfer commitment by an individual German investor firm is then derived from comparing the (relative) technological sophistication of a certain JV under observation with all of the (relative) sophistication levels of other joint ventures in this sample, controlling for a set of characteristics that vary between JVs.

That our assumption holds, i.e. that the maximum technological sophistication level of any joint ventures under observation is a result of technology transfer by the German firm, has been verified for the *average firm* by comparing the maximum sophistication levels of the German firm, the joint venture, and the Chinese partner. As the descriptive results provided in Section IV-1 have shown, the average technological sophistication of a joint venture exceeds that of the Chinese partner and can thus only be a result of technology transfer by the German investor. Additionally, one can verify the assumption *individually* by reading the qualitative descriptions of survey respondents regarding the roles and responsibilities of JV partners (see Appendix 6). For the joint ventures in this sample, there exists no described case in which sophisticated technological skills have been contributed by the Chinese partner.[117]

1.3. Ambiguity of Causal Chains

The third important aspect that we want to draw attention to is the ambiguity of causal chains. The object of investigation of this study is the current technological sophistication of a German-Chinese joint venture. This is used to infer the cooperative commitment regarding technology transfer by a German investor firm (see discussion above).

Consequently, the following concepts are used as a *dependent* variable in multiple regression analysis:

- the absolute sophistication of technological resources a JV is endowed with ($T(JV)$)
- its sophistication relative to headquarters [$T(JV)/T(HQ)$], and

---

[117] Unfortunately, two qualitative descriptions are missing. However, these two cases are not critical as the technology sophistication levels of the two respective Chinese partners are rather low.

- the degree to which sophisticated technology is shared with the JV partner.

Among the factors used as *independent* variables in multiple regression analysis are general factors, such as the characteristics of the German investor firm, the characteristics of technology employed by the joint venture, and the characteristics of the Chinese partner firm. Also, the variables derived from the theoretical hypotheses are used as independent variables.

As described in Section IV-2, the application of multiple regression analysis has produced statistically significant links between some of the independent variables and some of the dependent variables used. One could thus imply that there is a causal effect from the independent variable to the dependent variable. However, any *causality* cannot be proven, and some variables used in this study are vulnerable to the allegation of being endogenous.

The proof of causality from observational data such as survey results is generally problematic (Lieberson 1985, LaLonde 1986). As described by Winship and Morgan (1999), the basic problems that arise when using observational data are that (1) outcomes for the treatment and the control groups may differ even in the absence of the treatment and that (2) the potential effect of the treatment may differ for the treatment and control groups. The only way to remove these potential sources of bias is to use an appropriate experimental study design.

For this study, some of the observed effects can stem from mutually reinforcing effects or even reverse causality. One example of this is the influence of the *frequency of interaction* among partners on the degree of technological sophistication. The line of argumentation as derived for hypothesis 5b is that a high frequency of interaction allows a German JV partner to evaluate the Chinese JV partner's cooperative efforts and also to quickly detect uncooperative behaviour. This encourages the German partner to make commitments regarding technology transfer, because mutual cooperation is generally encouraged by strong reciprocity.

However, the opposite line of argumentation would also hold: because technologically sophisticated knowledge – especially tacit knowledge – can usually only be transferred via direct interaction of personnel, a high frequency of interaction increases the feasibility of knowledge transfer and thus results in a higher level of technological sophistication.

To list all possible reverse causal arguments for this study would go beyond the scope of this thesis. Instead we remind the reader that a significant relationship between an independent and a dependent variable does not necessarily claim the existence a causal effect from the respective independent variable to the dependent variable.

Besides the above-described limitations that affect our ability to interpret this study's results, there might exist a multitude of other restrictions. However, we do not elaborate on all potential caveats and restrict ourselves to the three main points described above for the sake of brevity. Next, we conclude our findings and give some suggestions for further research.

2. Conclusion and Suggestions for further Research

In recent years, economic growth rates and the large absolute size of the Chinese market have triggered a boom of German investment activity (Holtbrügge and Puck 2005, Kasperk et al. 2006). However, this investment boom has been accompanied by reports of damaging intellectual property diffusion (World Economic Forum 1995, Huck 2005, von Keller et al. 2005). As a result of the uncertainty regarding IP protection, foreign firms are deterred from deploying technological assets or capabilities to China or even shy away from market entry to China altogether (Geissbauer 1996, Gassmann and Zhen 2004).

Reservations of the foreign investor regarding technology diffusion can result in inter-firm goal incongruence when international market entry is accomplished by an international alliance. International joint ventures, the focus of this study, are regarded as a very suitable market entry form for transferring complex technological knowledge to a foreign country and are known to give a local partner the highest degree of insight (Root 1994, Inkpen and Beamish 1997). IJVs therefore not only represent a superb channel for German firms to transfer technology to China, but also for Chinese firms to absorb and misuse the absorbed IP (Tidd and Izumimoto 2002).

One thus realises the challenge for German industrial firms when deciding on how to endow international joint ventures in China with technological assets and capabilities. The importance of the technological assets at stake, the high likelihood of knowledge diffusion, and the element of goal incongruence within joint ventures add up to a complex decision framework.

This thesis investigated the market entry of German industrial firms to China by means of international joint ventures and the associated transfer of technology by German investors to the Chinese target market in order to shed light on this decision framework. Specifically, the questions that were investigated are:

What factors influence the sophistication of the technological endowment that an international joint venture in China receives from its German parent? and
In what way do strategic considerations regarding inter-firm cooperation and knowledge sharing influence the foreign investor's technology transfer behaviour?

Insights on these two questions have been generated by using theoretical frameworks on decision making and inter-organisational knowledge sharing to derive empirically testable hypotheses. These hypotheses were then tested using data from a survey among 34 German-Chinese joint ventures. The specific course of action can be summarised as follows.

In Chapter II, the relevant literature regarding the question at hand is reviewed, namely literature on technology, technological sophistication, technological leadership, international technology transfer and the specific circumstances of the target market, China. Also, the market entry form 'international joint venture' is reviewed with special attention to how it can be used as an effective channel for technology transfer.

Chapter III presents the application and adaptation of theoretical frameworks on cooperation and knowledge exchange to the case of technology transfer to international joint ventures. These theoretical considerations lead to the derivation of several testable hypotheses. Apart from our specific hypotheses, general factors that influence a foreign investor's endowment of an international joint venture with technological assets and capabilities are identified and described.

The design and the results of the empirical study are then presented in Chapter IV. Our qualitative and empirical results successfully establish a link between the co-operative setting of a German-Chinese joint venture and the sophistication of a joint venture's endowment with technological resources and capabilities. Empirical evidence for most of the previously derived hypotheses is found and discussed, leading to the overall acceptance of the view that arguments derived from the structural analysis of the cooperative setting within an international joint venture apply. Specifically:

- the cost structure of knowledge sharing for the German investor firm
- the level of mutual trust within the joint venture partnership
- the time horizon of the German investor and the age of a JV
- the frequency of interaction
- the behavioural transparency, and
- the mutual dependency of actors, taking into account the existence of multiple links among firms and the actors' degree of mutual dependency after the end of the specific JV contract at hand

have been statistically related to selected dimensions and measurement concepts of technological sophistication. This statistical evidence is summarised in Table 40 of Chapter IV, reproduced below:

Table 40: Overview of Statistical Results for Evaluation of Hypotheses (reproduced)

Hypotheses and relating Variables		Results for: T(JV) / [T(JV)/T(HQ)] / Shared_Technology				
(Predicted direction of effect in parentheses)		Technoware	Inforware	Humanware	Orgaware	Capabilities
Hypothesis 1a - d						
1a) Costs of own cooperation	C_{AC} (-)	n/n/n	n/n/n	-*/-**/n	n/n/n	n/n/n
1b) Costs of own defection	C_{AD} (+)	n/n/n	n/n/n	n/n/n	n/n/n	n/n/n
1c) Costs of being defected upon	C_{BD} (-)	n/n/n	n/n/n	n/n/n	n/n/n	n/n/n
1d) Synergy	V_S (+)	n/n/+**	n/n/+***	n/n/n	n/n/n	n/n/n
Hypothesis 2						
Trust	Trust (+)	n/n/n	n/n/n	n/n/n	n/n/n	+*/(+)/n
Hypothesis 3a & b						
3a) Age of JV	AgeJV (+)	n/n/n	n/n/n	+**/(+)/n	n/n/n	n/n/n
3b) Duration of Contract	LengthContract (+)	n/n/-*	n/n/n	n/n/n	n/n/n	n/n/n
Hypothesis 4a & b						
4a) Discount Rate	DiscountRate (-)	n/n/n	+**/+**/n	n/n/n	n/n/n	n/n/n
4b) Time Horizon	TimeHorizon (+)	n/n/n	n/n/n	n/n/n	+*/(+)/n	n/n/n
Hypothesis 5a & b						
5a) Frequency of Interaction	FreqInteract (+)	n/n/n	n/n/+**	n/n/(+)	n/n/n	n/n/+**
5b) Behavioural Transparency	Transparency (+)	+**/n/n	n/n/n	n/n/n	n/n/n	+*/+**/
Hypothesis 6a & b						
6a) Multiple Links	Links (+)	n/n/n	n/n/+**	n/n/n	n/n/n	n/n/+**
6b) Dependency	Dependency (+)	n/n/+*	n/n/+**	n/n/(+)	n/n/+*	n/n/n

*** significant at the 1% level   * significant at the 10% level   n = no significant result
** significant at the 5% level   ( ) weak support, see description in text

Source: Survey results.

The general picture offered by this overview is that survey results lend support to hypotheses on an individual basis, e.g. evidence supporting hypothesis 6a has been found in models explaining *Shared_Inforware* and *Shared_Capabilities*, but not in other models. Our survey thus offers a differentiated picture on individual relationships, but no broad and very robust support for one particular hypothesis.

Based on our research results, we provide the following suggestions for further research. Regarding the theoretical literature on the topic of cooperation in alliances, the author would encourage the development of comprehensive theoretical models that are applicable to knowledge sharing within organisations and take into account the factors that are empirically most relevant.

Second, improved empirical studies on the sophistication of a foreign activity's technological endowment could be designed. Future research could proceed by conducting a cross-country analysis, for example taking into account foreign investor firms from all EU countries or target markets from all across Asia.

Third, future studies could measure the degree of technological sophistication for a sample of different market entry forms, particularly joint ventures and WFOEs. This could help to identify the factors that increase an investor's commitment to invest technologically sophisticated resources or capabilities in a foreign country, irrespective of the market entry form. This approach would also enable researchers to compare the intra-group variability of technological sophistication with the inter-group variability of technological sophistication.

VI. References and Appendices

1. Reference List

Aharoni, Y. (1966). *The Foreign Investment Decision Process*, Graduate School of Business Administration, Harvard University, Boston.

Alchian, A. A. (1965). Some Economics of Property Rights, *Il Politico*, 30, p. 816-829.

Alchian, A. A.; Demsetz, H. (1973). The Property Rights Paradigm, *The Journal of Economic History*, 33, p. 16-27.

Al-Obaidi, Z. (1999). Modeling the International Technology Transfer (ITT) Process, in: Lehtinen, U., Seristoe, H. (Eds.), *Perspectives on Internationalization*, Helsinki School of Economics and Business Administration, Helsinki, p. 111-123.

Anderson, E.; Coughlan, A. T. (1987). International Market Entry and Expansion via Independent or Integrated Channels of Distribution, *Journal of Marketing*, Vol. 51, p.71-82.

Anderson, E.; Gatignon, H. (1986). Modes of Foreign Entry: A Transaction Cost Analysis and Propositions, *Journal of International Business Studies*, 17 (3), p. 1-26.

Andrews, K. R. (1971). *The Concept of Corporate Strategy*, Homewood, IL: R.D. Irwin.

Annand, B.; Khanna, T. (1997). *Intellectual Property Rights and Contract Structure*, Harvard Business School Working Paper, No. 97-016.

Argote, L. (1999). *Organizational Learning: Creating, Retaining, and Transferring Knowledge*, Norwell, MA: Kluwer.

© Springer Fachmedien Wiesbaden GmbH, part of Springer Nature 2008
M. Hoeck, *Cooperation and Technological Endowment in International Joint Ventures: German Firms in China*, Edition KWV, https://doi.org/10.1007/978-3-658-24355-5

Argote, L.; Ingram, P. (2000). Knowledge Transfer: A Basis for Competitive Advantage in Firms, *Organizational Behavior and Human Decision Processes*, 82 (1), p. 150- 169.

Arrow, K. (1969). Classificatory Notes on the Production and Transmission of Technological Knowledge, *American Economic Review*, Papers and Proceedings, 52 (May), p. 29-35.

Arrow, K.J. (1975). Gifts and Exchanges, in: Phelps, E.S. (Ed.), *Altruism, Morality, and Economic Theory*, New York, NY: Russell Sage Foundation, p. 13-28.

Axelrod, R. (1984). *The Evolution of Cooperation*, New York: Basic Books.

Axelrod, R.; Keohane, R. (1986). Achieving Cooperation under Anarchy: Strategies and Institutions, in: K. A. Oye (Ed.), *Cooperation under Anarchy*: p. 226-254, Princeton, NJ: Princeton University Press.

Ayal, I.; Zif, J. (1979). Market Expansion Strategies in Multinational Marketing, *Journal of Marketing*, 43, p. 84-94.

Bain, J.S. (1956). *Barriers to New Competition*, Cambridge, MA: Harvard University Press.

Ballhaus, J. (2005). Gegen Klau hilft nur die Offensive, in: *Absatzwirtschaft – Zeitschrift für Marketing* (5), p. 40.

Bamberger, I.; Wrona, T. (1996). Der Ressourcenansatz und seine Bedeutung für die Strategische Unternehmensführung, *Schmalenbachs Zeitschrift für betriebswirtschaftliche Forschung* (zfbf), 2/1996, p. 130-153.

Baranson, J. (1970). Technology Transfer Through the International Firm, *American Economic Review*, 60 (2), p. 435-440.

Barney, J. B. (1990). The Debate between Traditional Management Theory and Organizational Economics: Substantive Differences or Intergroup Conflict?, *Academy of Management Review*, 15, p. 382-393.

Barney, J. B. (1991). Firm Resources and Sustained Competitive Advantage, *Advances in Strategic Management*, Vol. 17, p. 203-227.

Bartlett, C. A.; Ghoshal, S. (1998), *Managing Across Borders: The Transnational Solution*, 2nd ed., Boston, MA: Harvard Business School Press.

Basu, K. (2007). The Traveler's Dilemma, *Scientific American*, 296 (6), p. 68-73.

Baum, J. A. C.; Ingram, P. (1998). Survival-Enhancing Learning in the Manhattan Hotel Industry, 1898-1980, *Management Science*, 44, p. 996-1016.

Beamish, P.W. (1985). The Characteristics of Joint Ventures in Developed and Developing Countries, *Columbia Journal of World Business*, 20 (3), p. 13-19.

Beckmann, C.; Fischer, J. (1994). Einflußfaktoren auf die Internationalisierung von Forschung und Entwicklung in der deutschen Chemischen und Pharmazeutischen Industrie, *Schmalenbachs Zeitschrift für betriebswirtschaftliche Forschung* (zfbf), 46, p. 630-657.

Benkard, C. L. (1999). *Learning and Forgetting: The Dynamics of Aircraft Production*, NBER Working Paper 7127, National Bureau of Economic Research, Cambridge, MA.

Bennett, D.; Liu, X.; Parker, D., Steward, F.; Vaidya, K. (2001). Technology Transfer to China: A Study of Strategy in 20 EU Industrial Companies, *International Journal for Technology Management*, 21, p. 151-176.

Bennett, D.J.; Vaidya, K.G.; Zhao, H.Y.; Wang, X.M. (1997). Transferring Manufacturing Technology to China: Supplier Perceptions and Acquirer Expectations, *The International Journal of Manufacturing Technology Management*, 8 (5), 283-91.

Birden, P.B. (1998). Technology Transfers to China: An Outline of Chinese Law, in: Tay, A.; Doeker-Mach, G. (Eds.), *Asia-Pacific Handbook*, Volume I, People's Republic of China.

Bleeke, J.; Ernst, D. (1991). The Way to Win in Cross-Border Alliances, *Harvard Business Review* 69 (Nov/Dec 1991), p. 127-135.

BMBF - Bundesministerium für Bildung und Forschung (2006). *Bericht zur Technologischen Leistungsfähigkeit Deutschlands 2006.*

Borys, B.; Jemison, D. B. (1989). Hybrid Arrangements as Strategic Alliances: Theoretical Issues in Organizational Combinations, *Academy of Management Review*, 14, p. 234-249.

Brandenburger, A.M.; Nalebuff, B.J. (1996). *Coopetition*, New York: Doubleday.

Bromiley, P.; Cummings, L. L. (1991). *Transaction Costs in Organizations with Trust*, Working Paper, University of Minnesota at Minneapolis.

Brouthers, L. E.; Brouthers, K. D., Werner, S. (1999). Is Dunning's Eclectic Framework Descriptive or Normative?, *Journal of International Business Studies*, 30 (4), p. 831-844.

Bruck, Jürgen (1996). *Entwicklung einer Gesamtkonzeption für das Management Strategischer Allianzen im F&E Bereich*, Schriften zur Unternehmensplanung, Band 37, Frankfurt am Main: Peter Lang GmbH.

Buckley, P.J. (1982). *Multinational Enterprises and Economic Analysis*, London: Cambridge University Press.

Buckley, P.J.; Casson, M. (1976). *The Future of the Multinational Enterprise*, London: Macmillan.

Buckley, P.J.; Casson, M. (1988). A Theory of Cooperation in International Business, in: Contractor, F. J.; Lorange, P. (Eds.). *Cooperative Strategies in International Business*. Lexington, MA: Lexington Books.

Buckley, P.J.; Casson, M. (1996). An Economic Model of International Joint Venture Strategy, *Journal of International Business Studies*, 27 (5), p. 849-876.

Buckley, P.J.; Newbould, G.D.; Thurwell, J.C. (1988). *Foreign Direct Investment by Smaller UK Firms: The Success and Failure of First-Time Investors Abroad*, 2nd edition, Basingstoke: Macmillan.

Büchel, B.; Prange, C.; Probst, G.; Rüling C.-C. (1998). *International Joint Venture Management. Learning to Cooperate and Cooperating to Learn.* Singapore: John Wiley & Sons.

Buzzell, R. D. (1968). Can You Standardize Multinational Marketing?, *Harvard Business Review*, 61, p. 92-101.

Caves, R.E. (1971). International Corporations: The Industrial Economics of Foreign Investment, *Economica*, 38 (1), p.1-27.

Caves, R.E. (1974). The Causes of Direct Investment: Foreign Firms' Shares in Canadian and UK Manufacturing Industries, *The Review of Economics and Statistics*, 56 (3), p. 279-293.

Caves, R.E.; Porter, M.E. (1977). From Entry Barriers to Mobility Barriers: Conjectural Decisions and Contrived Deterrence to New Competition, *Quarterly Journal of Economics*, 91, p. 241-262.

Chandler, A. D. (1962). *Strategy and Structure: Chapters in the History of the Industrial Enterprise*. Cambridge, MA: The MIT Press.

Chen, M. (1996), *Managing International Technology Transfer*, London: International Thomson Business Press.

China Statistical Yearbook 2000.

Chu, B. (1987). *Foreign Investment in China: A Question and Answer Guide*, Hong Kong: Hong Kong University Publisher and Printer.

Clifford, M.L.; Roberts, D.; Engardio, P. (1997). How You Can Win in China, *Business Week* (European Edition), 26.5.1997, p. 40-44.

Coase, R.H. (1937). The Nature of The Firm, in: Stigler, G.J.; Boulding, K.E. (Eds.), *Readings in Price Theory*, Homewood, IL: Irwin, p. 331-351.

Cohen, Goel (2004). *Technology Transfer: Strategic Management in Developing Countries*, New Delhi: Sage Publications.

Cohen, W. M.; Levinthal D. A. (1990). Absorptive Capacity: A New Learning Perspective on Learning and Innovation, *Administrative Science Quarterly*, 35, p. 128-152.

Collis, D. J. (1991). A Resource-Based Analysis of Global Competition: The Case of the Bearings Industry, *Strategic Management Journal*, 12, p. 49-68.

Conroy, R. (1992). *Technological Change in China*. OECD Development Center, Paris.

Contractor, F. J.; Lorange, P. (Eds.) (1988). *Cooperative Strategies in International Business*. Lexington, MA: Lexington Books.

Coughlin, C. C. (1983). The Relationship between Foreign Ownership and Technology Transfer, *Journal of Comparative Economics*, 7, p.400-414.

Currall, S. C.; Inkpen, A. C. (2002). A Multilevel Approach to Trust in Joint Ventures, *Journal of International Business Studies*, Vol. 33 (3), 479-495.

Daft, R. L.; Weicke, K. E. (1984). Toward a Model of Organizations as Interpretation Systems, *Academy of Management Review*, 9, p. 284-295.

Darr, E.; Argote, L.; Epple, D. (1995). The Acquisition, Transfer and Depreciation of Knowledge in Service Organizations: Productivity in Franchises, *Management Science*, 41, p. 1750-1762.

David, P. (1992). Knowledge, Property, and the Systems Dynamics of Technological Change, in: Summers, L.; Shah, A. (Eds.), *Proceedings of the World Bank Annual Conference on Development Economics 1992*, p. 215-248.

Davidson, W.H.; McFetridge, D.G. (1984). International Technology Transactions and the Theory of the Firm, *Journal of Industrial Economics*, 32 (3), p. 253-264.

Davidson, W.H.; McFetridge, D.G. (1985). Key Characteristics in the Choice of International Technology Transfer Mode, *Journal of International Business Studies*, 16 (2), p. 5-21.

de Almeida, P. R. (1995). The Political Economy of Intellectual Property Protection: Technological Protectionism and Transfer of Revenue among Nations, *International Journal of Technology Management*, 10 (2/3), p. 214-229.

Deutsche Bank Research (2005), *Offshoring-Report 2005 – Ready for Take-off*, Frankfurt.

Deutsche Bundesbank (2007). *Bestandserhebung über Direktinvestitionen*, Statistische Sonderveröffentlichung 10.

DG Bank, (1999). *China: Marktchance für den deutschen Mittelstand*, Frankfurt am Main

Dierickx, L; Cool, K. (1989). Asset Stock Accumulation and Sustainability of Competitive Advantage, *Management Science*, 35 (12), p.1504-1511.

Dodgson, M. (1993). Learning, Trust, and Technological Collaboration, *Human Relations*, 46(1), 77-96.

Dörrenbächer, H. (1992). Mythos Joint Venture: Ergebnisse einer Empirischen Untersuchung Deutsch-Sowjetischer Joint Ventures, *Osteuropa-Wirtschaft*, 37 (2), p.133-147.

Dong, H. (2004). *Intensivierung des Innovationsfördernden Technologie-transfers in China. unter besonderer Berücksichtigung der Reformpolitik*, Berichte aus der Volkswirtschaft, Aachen: Shaker Verlag.

Dony, A. G. C. (1999). *Market Entry Strategies for the PR China – An Empirical Study on the Beer and Soft Drink Industry*, PhD Thesis, Universität Wiesbaden.

Dore, R. (1984). Technical Self Reliance, in: Fransman, M.; King, K. (Eds.), *Technological Capability in the Third World*, London: Macmillan, p. 65-80.

Dosi G., Teece, D.; Winter, S. (1992). Toward a Theory of Corporate Coherence: Preliminary Remarks, in: Dosi, G.; Giannetti, R.; Toninelli, P.A. (Ed.), *Technology and the Enterprise in a Historical Perspective*, Oxford: Clarendon Press of Oxford University Press.

Dougherty, S. M. (1997). *The Impact of Technology Transfers on Industry Productivity in China: 1980-95*, MIT Science & Technology Initiative and International Trade Administration, Foreign Commercial Service, US Embassy in Beijing, China.

Doz, Y.; Prahalad, C. K.; Hamel, G. (1990). Control, Change, and Flexibility: The Dilemma of Transnational Collaboration, in: Bartlett, C.; Doz, Y.; Hedlund, G. (Eds.), *Managing the Global Firm*. London: Routledge, p. 117-143.

Doz, Y. (1996). The Evolution of Cooperation in Strategic Alliances: Initial Conditions or Learning Processes?, *Strategic Management Journal*, Special Issue, 17, p.55-83.

Dunning, J.H. (1977). Trade, Location of Economic Activity and the MNE: A Search for an Eclectic Approach, in: Ohlin, B.; Hesselborn, P.; Wijkman, M. (Eds.), *The International Allocation of Economic Activity*. New York: Holmes & Meier, p. 395-418.

Dunning, J.H. (1979). Explaining Changing Patterns of International Production: In Defence of the Eclectic Theory, in: *Oxford Bulletin of Economics and Statistics*, 41, p. 269-295.

Dunning, J.H. (1980). Toward an Eclectic Theory of International Production: Some Empirical Tests, *Journal of International Business Studies*, 11 (1), p. 9-31.

Dunning, J.H. (1981). *The Eclectic Theory of the MNC*, London: Allen & Unwin.

Dunning, J.H. (1993). *Multinational Enterprise and the Global Economy*, Workingham: Addison-Wesley.

Dunning, J.H.; Narula, R. (1995). The R&D Activities of Foreign Firms in the United States, *International Studies of Management & Organization*, 26, p. 461-491.

Dutta, P. K. (2000). *Strategies and Games: Theory and Practice*. Cambridge, MA: The MIT Press.

Dyer, G.; Maier, A. (2007). Danone droht Partner mit Prozess: Lebensmittelhersteller streitet mit chinesischem Konzern Wahaha über Zusammenarbeit, *Financial Times Deutschland*, 12.04.2007.

Eaton, J.; Kortum, S. (1996). Trade in Ideas: Patenting and Productivity in the OECD, *Journal of International Economics*, 40, p. 251-278.

Engelsman, E.C.; van Raan, A.F.J. (1994). A Patent Based Cartography of Technology, *Research Policy*, 23, p. 1-26.

Epple, D.; Argote, L.; Devadas, R. (1991). Organizational Learning Curves: A Method for Investigating Intra-Plant Transfer of Knowledge Acquired through Learning by Doing, *Organization Science*, 2 (1), p. 58-70.

Ernst, D.; O'Connor, D. (1989). *Technology and Global Competition: The Challenge for Newly Industrialising Economies*, OECD Development Centre Studies, Paris.

European Commission (2003). Article 62 Decision, case No COMP/M.3149 - PROCTER & GAMBLE / WELLA, SG (2003) D/231120, Brussels, 30.07.2003.

Feess, E. (2004). *Mikroökonomie: Eine spieltheoretische- und anwendungsorientierte Darstellung*, 3rd Edition. Marburg, Metropolis.

Feess, E., Hoeck, M.; Lorz, O. (2006), *Oligopolistic Competition and International Technology Transfers*, Working Paper, RWTH Aachen University.

Frame, J. D. (1983). *International Business and Global Technology*, Lexington, MA: Heath and Company.

Fransman, M. (1984). *Technological Capability in the Third World*, London: Macmillan.

Frederiksen, L.; Sedita, S. R. (2005). *Embodied Knowledge Transfer: Comparing Inter-Firm Labor Mobility in the Music Industry and Manufacturing Industries*. DRUID Working Paper No. 2005-14.

Freeman, R. E. (1987). Review of 'The Economic Institutions of Capitalism' by 0. E. Williamson, *Academy of Management Review*, 12, p.385-387.

Freeman, C. (1990). Networks of Innovators: A Synthesis of Research Issues, *Research Policy*, 20, p. 499-514.

Freericks, C. (1998). *Internationale Direktinvestitionen mittelständischer Unternehmen*, Berlin: VWF Verlag für Wissenschaft und Forschung.

Friedmann, J. (1971). A Non-Cooperative Equilibrium for Supergames, *Review of Economic Studies,* 38, p. 1-12.

Fu, G. (2005). *Internationale Markteintrittsstrategien mittelgroßer Industrie-unternehmen (MU) am Beispiel deutscher Unternehmen in China.* Dissertation. Universität Stuttgart.

Gärtner, M. (2004). Markenimitat auf vier Rädern, *Handelsblatt*, 20.02.2004.

Gahl, A. (1991). *Die Konzeption strategischer Allianzen*, Dissertation, Universität Münster.

Gassmann, O.; Zhen, H. (2004). Motivations and Barriers of Foreign R&D Activities in China, *R&D Management*, 34 (4), p. 423-437.

Geissbauer, R. (1996). Direktinvestitionen in der VR China: Drahtseilakt zwischen Chaos und Erfolg, in: Gößl, M. M.; Lemper, A. (Eds.), *Geschäftspartner VR China*, Münster: LIT Verlag, p. 121-146.

Geringer, J. M.; Hebert, L. (1991). Measuring Performance of International Joint Ventures, *Journal of International Business Studies*, 22, p. 249-264.

Gersemann, O.; Hohensee, M.; Köhler, A.; Schnaas, D.; Sieren, F. (2002). Aus dem Boden gestampft, *Wirtschaftswoche*, (46), 2002.

Gibbons, R. (1992a), *A Primer in Game Theory*, London: Prentice Hall.

Gibbons, R. (1992b), *Game Theory for Applied Economists*, Princeton, NJ: Princeton University Press.

Glück, U. (2001). Entwicklung von Zentral- und Westchina: Neue Durchführungsbestimmungen, *China Kontakt*, March 2001, p.35.

Goh, A.-T. (2004). *Knowledge Diffusion, Supplier's Technological Effort and Technology Transfer via Vertical Relationships*. CEPR Discussion Paper No 4085.

Granovetter, M. (1985). Economic Action and Social Structure: The Problem of Embeddedness, *American Journal of Sociology*, 78, p. 481-510.

Granstrand, O., Hakanson, L.; Sjölander, S. (1992). *Technology Management and International Business: Internationalization of R&D and Technology*, Chichester: Wiley.

Granstrand, O.; Hakanson, L.; Sjölander, S. (1993). Internationalization of R&D – A Survey of Some Recent Research, *Research Policy*, 22, p. 413-430.

Gray, B. (1989). *Collaborating*, San Francisco: Jossey-Bass.

Greeven, M. J. (2004). *The Evolution of High-Technology in China after 1978: Towards Technological Entrepreneurship*. ERIM Report Series Reference No. ERS-2004-092-ORG.

Grubel, H. G. (1967). Intra-Industry Specialisation and the Patterns of Trade, *The Canadian Journal of Economic and Political Science*, 33, p. 374-388.

Gulati, R.; Khanna, T.; Nohria, N. (1994). Unilateral Commitments and the Importance of Process in Alliances, *Sloan Management Review*, 35 (3), p. 61-69.

Gulati, R. (1995). Does Familiarity Breed Trust? The Implications of Repeated Ties for Contractual Choice in Alliances, *Academy of Management Journal*, 38, p. 85-112.

Gulati, R.; Singh, H. (1998). The Architecture of Cooperation: Managing Coordination Costs and Appropriation Concerns in Strategic Alliances, *Administrative Science Quarterly*, Volume 43 (4), p. 781-814.

Gutterman, A. S. (1997). *International Joint Ventures. Negotiation, Formation and Operation*, Nestor House: Euromoney.

Hagedoorn, J. (1993). Understanding the Rationale of Strategic Technology Partnering: Interorganizational Modes of Cooperation and Sectoral Differences, *Strategic Management Journal*, 14, p. 371-385.

Hakanson, L. Nobel, R. (1993). Foreign R&D by Swedish Multinationals, *Research Policy*, 22, p. 373-396.

Hamel, G. (1991). Competition for Competence and Inter-Partner Learning within International Strategic Alliances, *Strategic Management Journal*,12, p. 83-104.

Hamel, G.; Doz, Y. L.; Prahalad, C. K. (1989). Collaborate with your Competitors – and Win, *Harvard Business Review*. 67(1), p. 133-139.

Hamel, G.; Prahalad, C. K. (1990). The Core Competence of the Corporation, in: *Harvard Business Review*, May-June 1990, p. 79-81.

Hardy, M.; Bryman, A. (2004). *Handbook of Data Analysis*, London: SAGE Publications.

Harrigan, K. R. (1986). *Managing for Joint Venture Success*, New York: Lexington Books.

Harrigan, K. R.; Newman, W. H. (1990). Bases of Interorganizational Cooperation: Propensity, Power, Persistence, *Journal of Management Studies*, 27, p. 417-434.

Heckscher, E. F. (1919). Utrikehandels verkam pa inkomst – fördelingen. *Ekonomisk Tidskrift*, 21, p. 497-512. Reprinted in 1950 as: The Effect of Foreign Trade on the Distribution of Income, in: Ellis, H. S.; Metzler, L. A. (Eds.), *Readings in the Theory of International Trade*, London: Allen & Unwin, p. 272-300.

Hedlung, G.; Kverneland, A. (1985). Are Strategies for Foreign Markets Entry Modes Changing: The Case of Swedish Investment in Japan, *International Studies of Management and Organization*, 15. (2), p. 41-59.

Heide, J. B.; Miner, A. S. (1992). The Shadow of the Future: Effects of Anticipated Interaction and Frequency of Contact on Buyer-Seller Cooperation, *Academy of Management Journal*, 35, p. 265-291.

Heidhues, F. (1969). *Zur Theorie der internationalen Kapitalbewegungen – Eine kritische Untersuchung unter besonderer Berücksichtigung der Direktinvestitionen*, Tübingen: J.C.B. Mohr.

Hennart, J. F. (1982). *A Theory of Multinational Enterprise*, Ann Arbor, MI: University of Michigan Press.

Hennart, J. F. (1988). A Transaction Cost Theory of Equity Joint Ventures, *Strategic Management Journal*, 9, p. 361-374.

Hennart, J. F. (1991). The Transaction Cost Theory of Joint Ventures, *Management Science*, 37, p. 483-497.

Hennart, J. F.; Roehl, T.; Zietlow, D. S. (1995). *"Trojan horse" or "workhorse"? The evolution of U.S.-Japanese joint ventures in the United States-Revised*. CIBER Working paper 95-103, College of Commerce and Business Administration, University of Illinois, Urbana-Champaign, IL.

Hermann, R. (1988). *Joint Venture Management: Strategien, Strukturen, Systeme und Kulturen*, Dissertation, St.Gallen.

Heuser, R.; Klein, R. (Eds.) (2004). *Die WTO und das neue Ausländer-investitions- und Außenhandelsrecht der VR China – Gesetze und Analysen*. Hamburg.

Hildebrandt, L.; Weiss, C. (1997). Internationale Markteintrittsstrategien und der Transfer von Marketing-Know-how, *Zeitschrift für betriebswirtschaftliche Forschung*, 49 (1), p. 3-25.

Hilger, A. (2001). *Erfolgsfaktoren für Internationalisierungsstrategien*, Frankfurt a. Main: Peter Lang.

Hill, C. W. (1990). Cooperation, Opportunism, and the Invisible Hand: Implications for Transaction Cost Theory, *Academy of Management Review*, 15, p. 500-513.

Hill, C. W.; Kim, W.C. (1988). Searching for a Dynamic Theory of the MNE: A Transaction Cost Approach, *Strategic Management Journal*, 9, p. 93-104.

Hladik, K.J. (1988). R&D and International Joint Ventures, in: Contractor, F.J.; Lorange, P. (Eds.). *Cooperative Strategies in International Business – Joint Ventures and Technology Partnerships between Firms*, Lexington: Lexington Books, p.188-203.

Hobday, M. (1995). *Innovation in East Asia: The Challenge to Japan*. Aldershot: Edward Elgar.

Hofstede, G. (1980). *Culture's Consequences: International Differences in Work-Related Values*, Newbury Park, CA: Sage Publications.

Hofstede, G. (2003). *Culture's Consequences, Comparing Values, Behaviors, Institutions, and Organizations across Nations.* Second Edition, Newbury Park, CA: Sage Publications.

Holtbrügge, D.; Puck, J.F. (2005). *Geschäftserfolg in China: Strategien für den größten Markt der Welt*, Berlin: Springer Verlag.

Horsley, J.P. ; Sullivan, E. (1995). *East Asian Executive Reports*, International Executive Reports Ltd.

Huber, G. P. (1991). Organizational Learning: The Contributing Processes and a Review of the Literatures, *Organization Science,* 2, p. 88-117.

Huck, W. (2005). *Rahmenbedingungen für den Technologietransfer nach China*, Braunschweig: IBL.

Hymer, S. (1960). The International Operations of National Firms: A Study of Direct Investment, PhD Thesis, MIT, Cambridge, MA.

Hymer, S. (1976). The International Operations of National Firms: A Study of Direct Investment. Cambridge, MA: MIT Press: (reprint of Ph.D. Thesis).

IBM Business Consulting (2006). *Inside China – How the Chinese view their automotive future*. Report.

Inkpen, A. C.; Beamish, P. W. (1997). Knowledge, Bargaining Power, and the Instability of International Joint Ventures, *The Academy of Management Review*, 22 (1), p. 177-202.

Inkpen, A. C. (1995). *The Management of International Joint Ventures: An Organizational Learning Perspective*, London: Routledge.

Inkpen, A. C.; Crossan, M. M. (1995). Believing is Seeing: Joint Ventures and Organization Learning, *Journal of Management Studies*, 32, p. 595-618.

ITEM/TECTEM (2005). *Internationalization of the Value Chain*, Benchmarking project report, University of St. Gallen.

Jarillo, J. C. (1988). On Strategic Networks, *Strategic Management Journal*, 9 (1) , p. 31-42.

Johanson, J.; Valhne, J.-E. (1977). The Internationalization Process of the Firm – A Model of Knowledge Development and Increasing Foreign Market Commitments, *Journal of International Business Studies*, p. 23-32.

Johanson, J.; Wiedersheim-Paul, F. (1975). The Internationalization of the firm – Four Swedish Cases. *Journal of Management Studies*, 12 (3), p. 305-360.

Johanson, J.; Vahlne, J.-E. (1990). The Mechanism of Internationalisation, *International Marketing Review*, 7 (4), p. 11-24.

Johnson, H.G. (1970). The Efficiency and Welfare Implications of the International Corporation, in: Kindleberger, Ch. (Ed.), *The International Corporation*, Cambridge, MA: MIT Press, p. 35-56.

Kabiraj, T.; Marjit, S. (1992). To Transfer or Not to Transfer the Best Technology under Threat of Entry – the Case of Price Competition, in: Dutta, B.; Mookherjee, D.; Parthasarathy, T.; Raghavan, T.; Ray, D.; Tijs, S. (Eds.), *Game Theory and Economic Applications*, p. 356-368.

Kahen, G. (1994). *A Comprehensive and Strategic Model of Technology Transfer: Emphasizing IT*, IS UK PhD Consortium, Cranfield, UK.

Kahen, G.; Griffiths, C. (1995). Human Factors, Technology Transfer, and Information Technology in the Socio-Economic Development of the Third World, *Proceedings of IT-DEV Conference 1995*, University of Johannesburg, p. 188-206.

Kale, P.; Singh, H.; Perlmutter, H. (2000). Learning and Protection of Proprietary Assets in Strategic Alliances: Building Relational Capital, *Strategic Management Journal*, 21 (3), p. 217-237.

Kasperk, G. Woywode M.; Kalmbach R. (2006). *Erfolgreich in China. Strategien für die Automobilzulieferindustrie*, Berlin: Springer Verlag.

Katz, R.; Rebentisch, E. S.; Alien, T.J. (1996), A Study of Technology Transfer in a Multinational Cooperative Joint Venture, *IEEE Transactions on Engineering Management*, 43 (1), p. 97-105.

Kaufmann, F. (1995). Internationalisierung Mittelständischer Unternehmen, *Die Unternehmung*, 3/1995, p. 201-213.

Keller, W. (1996). Absorptive Capacity: On the Creation and Acquisition of Technology in Development, *Journal of Development Economics,* 49, p. 199-227.

Keller, W. (2004). International Technology Diffusion, *Journal of Economic Literature*, 42(3), p. 752-782.

Kenen, P. B. (1965). Nature, Capital and Trade, *Journal of Political Economy*, 73, p. 437-460.

Killing, J. P. (1983). *Strategies for Joint Venture Success*, New York: Praeger.

Kindleberger, Ch. P. (1969). *American Business Abroad*, New Haven: Yale Press.

Klemperer, P. (1999). Auction Theory: A Guide to the Literature, *Journal of Economic Surveys*, 13, p. 227-286.

Kloth, B. (1996). *Technologietransfer im Rahmen Schweizerisch-Chinesischer Joint Ventures*, PhD Thesis, St.Gallen University.

Knickerbocker, F.T. (1973). *Oligopolistic Reaction and Multinational Enterprise*, Cambridge, MA: Harvard University Press.

Knight, F. H. (1965). *Risk, Uncertainty and Profit*, New York: Harper & Row.

Kögel, P.; Gälli, A. (1997). Großchina auf dem Weg zum High-Tech-Standort?, *ifo Forschungsberichte der Abteilung Entwicklungsländer*, Köln.

König, M. (2003). An Econometric Framework for Testing the Eclectic Paradigm of International Firm Activities, *Review of World Economics*, 139 (3) , p. 484-506.

Kogut, B. (1988a). A Study of the Life Cycle of Joint Ventures, *Management International Review*, 28, p. 39-52.

Kogut, B. (1988b). Joint Ventures: Theoretical and Empirical Perspectives, *Strategic Management Journal*, 9, p. 319-322.

Kogut, B.; Chang, S. (1991). Technological Capabilities and Japanese Foreign Direct Investment in the United States, *Review of Economics and Statistics*, 73, p. 401-413.

Kogut, B.; Zander, U. (1993). Knowledge of the Firm and Evolutionary Theory of the Multinational Corporation, *Journal of International Business Studies*, 24, p. 625-645.

Koller, H.; Raithel, U.; Wagner, E. (1998). Internationalisierungsstrategien Mittlerer Industrie-Unternehmen am Standort Deutschland, *Zeitschrift für Betriebswirtschaft*, 68 (2), p. 175-203.

Konrad, R. (1989). *Chancen und Risiken von Equity Joint Ventures in der Volksrepublik China*, PhD Thesis, Universität St. Gallen.

Kotabe, M. (1989). 'Hollowing-out' of U.S. Multinationals and their Global Competitiveness, *Journal of Business Research*, 19 (August), p. 1-15.

Kotabe, M. (1992), *Global Sourcing Strategy*, New York: Quorum.

Kravis, I. B. (1956). Availability and Other Influences on the Commodity Composition of Trade, *Journal of Political Economy*, 64 (2), p. 143-155.

Kumar, B.N. (1989). Formen der Internationalen Unternehmenstätigkeit, in: Macharzina, K.; Welge, M.K. (Eds.), *Handwörterbuch Export und Internationale Unternehmung*, Stuttgart, p. 914-926.

Kutschker, M. (1997). Markteintrittsformen in China, in: M. Kutschker (Ed.), *Management in China*, Frankfurt am Main, p. 65-86.

Kutschker, M. (1999). Das Internationale Unternehmen, in: Kutschker, M. (Ed.). *Perspektiven der internationalen Wirtschaft*, Wiesbaden, p.101-125.

Kutschker, M.; Schmid, S. (2002). *Internationales Management*, München: Oldenbourg.

LaLonde, R.J. (1986). Evaluating the Econometric Evaluations of Training Programs with Experimental Data, *American Economic Review*, Volume 76, p. 604-620.

Langhauser, S. (2000). *Marktchancenanalyse – Theoretischer Teil*, German Chamber of Commerce in China, Beijing.

Leontief, W. W. (1956). Factor Proportions and the Structure of American Trade: Further Theoretical and Empirical Analysis, *Review of Economics and Statistics*, 28, p. 368-390.

Levin, R.; Klevorick, A.; Nelson, R.; Winter, S. (1987). Appropriating the Returns from Industrial Research and Development, *Brookings Papers on Economic Activity,* 3, p.783-820.

Levitt, T. (1983). Globalisation of Markets, *Harvard Business Review*, 61, p. 92-102.

Li, J.; Zhong, J. (2003). Explaining the Growth of International R&D Alliances in China, *Managerial and Decision Economics*, 24, p. 101-105.

Lieberson, S. (1985). *Making it Count: The Improvement of Social Research and Theory*, Berkeley: University of California Press.

Lin, Z. (2000). *Standortwahl deutscher Mittelständler in der VR China und Maßnahmen chinesischen Standortmarketings aus spieltheoretischer Sicht*, PhD Thesis, Berlin.

Linder, S. B. (1961). *An Essay on Trade and Transformation*, New York: John Wiley.

Lippman, S.; Rumelt, R.P. (1982). Uncertain Imitability: An Analysis of Inter-Firm Differences in Efficiency under Competition, *Bell Journal of Economics*, 13 (2), p. 418-438.

Litter, D. (1988). *Technological Development*, Oxford: Philip Allen.

Loebecke, C.; van Fenema, P.; Powell, P. (1998). Knowledge Transfer under Co-opetition, in: Larsen, T.; Levine, L.; DeGross, J. (Eds.), *Information Systems: Current Issues and Future Changes*, Laxenburg, Austria, p. 215-229.

Macaulay, S. (1963). Non-Contractual Relations in Business: A Preliminary Study, *American Sociological Review*, 28(1), p. 55- 67.

Macharzina, K. (1994). Joint Ventures, in: Dülfer, E. (Ed.), *International Handbook of Cooperative Organizations*, Göttingen, p.522-527.

Macharzina, K. (2003). *Unternehmensführung: Das internationale Managementwissen: Konzept - Methoden – Praxis*, Fourth edition, Wiesbaden: Gabler Verlag.

Mankiw, N.G. (1997), *Macroeconomics*, Third Edition, New York: Worth Publishers.

Mansfield, E.; Romeo, A. (1980). Technology Transfer through Overseas Subsidiaries by U.S.-based Firms, *Quarterly Journal of Economics*, 94, p. 737-750.

Mansfield, E.; Romeo, A.; Wagner, S. (1979). Foreign Trade and U.S. Research Development, *The Review of Economics and Statistics*, 61, p. 49-57.

Mansfield, E. (1995). *Intellectual Property Protection, Direct Investment, and Technology Transfer. Germany, Japan, and the United States*, International Finance Corporation Discussion Paper, The World Bank, Washington, D.C.

Margin, V.; Schunk, H., Heil, O.; Fürst, R. (2003). Kooperation und Coopetition: Erklärungsperspektive der Spieltheorie, in: Zentes, J., Bernhard S.; Morschett, D. (Eds), *Kooperationen, Allianzen und Netzwerke*, Wiesbaden: Gabler.

Markusen, J. R. (1995). The Boundaries of Multinational Enterprises and the Theory of International Trade, *Journal of Economic Perspectives*, 9, p. 169-190.

Martin, A. (1989). *Die empirische Forschung in der Betriebswirtschaftslehre*, Suttgart: Schäffer-Poeschel Verlag.

Maskus, Keith E. (2000). *Intellectual Property Rights in the Global Economy*, Institute for International Economics: Washington D.C.

Maskus, Keith E. (2003). *Encouraging International Technology Transfer*, University of Colorado (UNCTAD/ICTSD Capacity Building Project).

McManus, J. C. (1972). The Theory of the International Firm, in: Paquet, G. (Ed.), *The Multinational Firm and the Nation State*, Toronto: Collier-Macmillan, p. 66-93.

Meffert, H. (1998). *Marketing*, 8th Edition, Wiesbaden.

Merges, R.; Nelson, R. (1990). On the Complex Economics of Patent Scope, *Columbia Law Review*, 90, p. 839-870.

MOFCOM (2004). Internet source. Last accessed at: http://tfs.mofcom.gov.cn/aarticle/Zcfb/20050300029076.html on December 12th, 2005.

MOFCOM (2005a). Internet source. Last accessed at http://kjs.mofcom.gov.cn/aarticle/bn /cbw/200512/20051201045062.html on December 12th, 2005.

Mohr, A.T.; Puck, J.F. (2003). *Exploring the Determinants of the Trust-Control Relationship in International Joint Ventures*, Bradford University School of Management, Working Paper 3/24, Bradford.

Nash, J. F. (1950). Equilibrium Points in N-Person Games, *Proceedings of the National Academy of Sciences of the United States of America*, 36, p. 48-49.

Nelson, R. R.; Winter, S. G. (1982). *An Evolutionary Theory of Economic Change*. Cambridge, MA: Harvard University Press.

Nelson, R. R.; Phelps, E. S. (1966). Investments in Humans, Technological Diffusion and Economic Growth, *American Economic Review*, Papers and Proceedings 56, p. 69-75.

Niosi, J.; Hamel, P.; Fiset, L. (1995). Technology Transfer to Developing Countries through Engineering Firms: The Canadian Experience, *World Development*, 23 (10), p. 1815-1824.

Nolte, R.; Schrick-Hildebrand, P. (2004). *Deutscher Mittelstand in China – Status quo und Perspektiven*. Study of the German IKB Bank, Düsseldorf.

Nonaka, I. (1991). The Knowledge-Creating Company. *Harvard Business Review*, 69, p. 96-104.

Nonaka, I. (1994). A Dynamic Theory of Organizational Knowledge, *Organization Science*, 5, p.14-37.

Nonaka, I.; Takeuchi, H. (1995). *The Knowledge-Creating Company: How Japanese Companies create the Dynamics of Innovation*, New York: Oxford University Press.

Nurkse, R. (1934). Ursachen und Wirkung internationaler Kapitalbewegungen, *Zeitschrift für Nationalökonomie*, 5, p. 78-96.

Odedra, M. R. (1990). *The Transfer of Information Technology to Development Countries: Case Studies from Kenya, Zambia, and Zimbawe*, PhD Thesis, London School of Economics, London.

OECD (1998). *Internationalization of Industrial R&D: Pattern and Trends*, Paris.

Ohlin, B. (1933). *Interregional and International Trade*, Cambridge, MA: Harvard University Press.

Ohmae, K. (1985). *Macht der Triade – Die neue Form weltweiten Wettbewerbs*, Wiesbaden: Gabler.

Oliver, C. (1990). Determinants of Interorganizational Relationships: Integration and Future Directions, *Academy of Management Review*, 15, p. 241-265.

Ono, Y. (1991). Borden's Breakup with Meiji Milk shows how a Japanese Partnership can Curdle, *Wall Street Journal,* February 21, p. B1 & B6.

Osborn, R. N.; Baughn, C. C. (1990). Forms of Interorganizational Governance for Multinational Alliances, *Academy of Management Journal*, 33, p. 503-519.

Osborne, M. J.; Rubinstein, A. (1994), *A Course in Game Theory*, Cambridge, MA: The MIT Press.

Oster, A. (1990). *Modern Competitive Analysis*, New York: Oxford University Press.

Oxford English Dictionary, Fourth Edition, Oxford: Oxford University Press.

Oxley, J. E. (1997). Appropriability Hazards and Governance in Strategic Alliances: A Transaction Cost Approach, *Journal of Law, Economics, and Organization*, 13, p. 387-409.

Oye, K. A. (1986). *Cooperation under Anarchy*, Princeton, NJ: Princeton University Press.

Panhans, D.; Dingeldein, R. (2005). *International Expansion Strategies. Strategies for China - Value Creation Structures.* McKinsey Report.

Paquin, R. (2000). *Internationale Joint Venture als Organisationsform des Technologietransfers*, Berlin: Duncker & Humblot.

Parkhe, A. (1991). Interfirm Diversity, Organizational Learning, and Longevity in Global Strategic Alliances, *Journal of International Business Studies*, 22, p. 579-602.

Parkhe, A. (1993). Strategic Alliance Structuring: A Game Theoretic and Transaction Cost Examination of Interfirm Cooperation, *Academy of Management Journal*, 36, p. 794-829.

Pausenberger, E. (1982). Technologiepolitik internationaler Unternehmen, *Schmalenbachs Zeitschrift für betriebswirtschaftliche Forschung* (zfbf), 34 (12), p. 1025-1054.

Perlmutter, H. V. (1969). The Tortuous Evolution of the Multinational Corporation, *Columbia Journal of World Business*, 4 (Jan-Feb), p. 9-18.

Pisano, G. P. (1989). Using Equity Participation to Support Exchange: Evidence from the Biotechnology Industry, *Journal of Law, Economics, and Organization,* 5, p.109-126.

Pisano, G. P.; Russo, M.V.; Teece, D. (1988). Joint Ventures and Collaborative Agreements in the Telecommunication Industry, in: Mowery, D. (Ed.), *International Collaborative Ventures in U.S. Manufacturing*, Cambridge, MA: Ballinger, p. 23-70.

Polanyi, M. (1958). *Personal Knowledge: Towards a Post-Critical Philosophy*, Chicago, IL: University of Chicago Press.

Porter, M. E. (1980). *Competitive Strategy: Techniques for Analyzing Industries and Competitors.* New York: Free Press.

Porter, M. E. (1981). The Contributions of Industrial Organization to Strategic Management, *Academy of Management Review*, 6, p. 609-620.

Porter, M. E. (1985). *Competitive Advantage: Creating and Sustaining Superior Performance*, New York: Free Press.

Porter, M. E. (1986). *Competition in Global Industries*, Boston: Harvard Business School Press.

Porter, M. E. (1990). *The Competitive Advantage of Nations*, New York: Free Press.

Porter, M. E. (1991). *Nationale Wettbewerbsvorteile. Erfolgreich konkurrieren auf dem Weltmarkt*, München.

Posner, M. V. (1961). Technical Change and International Trade, *Oxford Economic Papers*, 13, p. 323-341.

Powell, W. W. (1990). Neither Market nor Hierarchy: Network Forms of Organization. *Research in Organizational Behavior*, 12, p. 295-336.

Prahalad, C. K. (1975). *The Strategic Process in a Multinational Corporation*, PhD Thesis, Harvard Graduate School of Business Administration.

Pucik, V. (1991). Technology Transfer in Strategic Alliances, Competitive Collaboration and Organizational Learning, in: Agmon, T.; von Glinow, M. A. (Eds.), *Technology Transfer in International Business*, New York: Oxford University Press, p. 121-138.

Ramanathan, K., (1994). Technology Choice: An Integrated Approach for the Choice of Appropriate Technology, *Science and Public Policy*, 21 (4), p. 221-232.

Rapoport, A.; Chammah, A. (1965). *Prisoner's Dilemma*, University of Michigan, Ann Arbor.

Reich, R. B.; Mankin, E. D. (1986). Joint Ventures with Japan give away our Future, *Harvard Business Review*, 64 (2), p. 78-86.

Ricardo, D. (1817). *Principles of Political Economy and Taxation*, London: Bell.

Rieck, C. (1993). *Spieltheorie: Einführung für Wirschafts- und Sozialwissenschaftler*, Wiesbaden: Gabler.

Ring, P. S.; van de Ven, A. (1992). Structuring Cooperative Relationships between Organizations, *Strategic Management Journal*, 13 (7), p. 483-498.

Ring, P. S.; van de Ven, A. (1994). Developmental Processes of Cooperative Interorganizational Relationships, *Academy of Management Review*. 19, p. 90-118.

Rockett, K. (1990). The Quality of Licensed Technology, *International Journal of Industrial Organization*, 8, p. 559–574.

Roehrig, M. F. (1994). *Foreign Joint Ventures in Contemporary China*, London: MacMillan.

Rogers, E. M. (2003). *Diffusion of Innovations*, Fifth Edition. New York: Free Press.

Root, F. R. (1994). *Entry Strategies for International Markets*. New York: Lexington.

Ross, S. A.; Westerfield, R. W.; Jaffe, J. (1999). *Corporate Finance*. Fifth Edition, Boston: Irwin McGraw-Hill.

Rouach, D. (2003). Technology Transfer and Management: Guidance for Small and Medium-Sized Enterprises, *Tech Monitor*, May-June 2003, p.25-28.

Rugman, A. M. (1975). Motives of Foreign Direct Investment: The Market Imperfections and Risk Diversification Hypothesis, *Journal of World Trade Law*, 9 (5), p. 567-573.

Rumelt, R. (1984). Towards a Strategic Theory of the Firm, in: Lamb, R. (Ed.), *Competitive Strategic Management*, Englewood Cliffs, NJ: Prentice-Hall, p. 556-570.

Saha, G. C.; Islam, N. (1998), Technological Information for Technology Strategy Management, *International Journal of the Computer, the Internet and Management*, 6 (3), p. 28-44.

Samuelson, P. A. (1954). The Pure Theory of Public Expenditure, *Review of Economics and Statistics*, 36 (4), p. 387-389.

Schelling, T. C. (1956). An Essay on Bargaining, *American Economic Review*, 46, p. 281-306.

Schelling, T. C. (1960). *The Strategy of Conflict*, Cambridge, M.A: Harvard University Press.

Scherer, F. M. (1975). The Determinants of Multi-Plant Operation in Six Nations and Twelve Industries, *Kyklos*, 27, p. 124-139.

Scherer, F. M. (1980). *Industrial Market Structure and Economic Performance*, Second Edition, Boston: Hough-Mifflin.

Schierenbeck, H. (1995). *Grundzüge der Betriebswirtschaft*, München.

Schrader, S. (1990). *Zwischenbetrieblicher Informationstransfer. Eine Empirische Analyse Kooperativen Verhaltens*, Berlin: Dunker & Humbolt.

Schwödiauer, G. (1992). Spieltheorie, in: Gal, T. (Ed.), *Grundlagen des Operations Research*, Third Edition, Berlin, p. 1-68.

Seabright, M. A., Levinthal, D. A.; Fichman, M. (1992). Role of Individual Attachments in the Dissolution of Interorganizational Relationships, *Academy of Management Journal*, 35, p. 122-160.

Seitz, K. (2002). *China – Eine Weltmacht kehrt zurück*, Berlin: Berliner Taschenbuch.

Sharif, N. (1986). *Technology Policy Formulation and Planning: A Reference Manual*, Asian and Pacific Center for Transfer of Technology, Bangalore.

Sharif, N. (1988). Basis for Techno-Economic Policy, *Science and Public Policy*, 15 (4), p. 217-229.

Sharif, N. (1994). Integrating Business and Technology Strategies in Developing Countries, *Technological Forecasting and Social Change*, 45, p. 151-167.

Sharif, N. (1995). The Evolution of Technology Management Studies: Technoeconomics to Technometrics, *Technology Management*, 2, p. 113-148.

Sharif, N. (1997). Technology Strategy in Developing Countries: Evolving from Comparative to Competitive Advantage, *International Journal of Technology Management*, 10 (10), p. 1-33.

Shi, H. (1994). *Wirtschaftskooperationen und Technologietransfer in China mit besonderer Berücksichtigung von Equity Joint Ventures*, PhD Thesis, Universität Wuppertal.

Simon, H. (1996). *Die Heimlichen Gewinner (Hidden Champions). Die Erfolgs-strategien unbekannter Weltmarktführer*, Frankfurt am Main.

Simonin, B. L.; Helleloid, D. (1993). Do Organizations Learn? An Empirical Test of Organizational Learning in International Strategic Alliances, in: Moore, D. (Ed.), *Academy of Management Best Paper Proceedings*, p. 222-226.

Sinha, U. B. (2001). International Joint Venture, Licensing and Buy-out under Asymmetric Information, *Journal of Development Economics*, 66, p. 127-151.

Smith, A. (1776). *An Inquiry into the Nature and Courses of the Wealth of Nations*, London.

Solow, R. M. (1956). A Contribution to the Theory of Economic Growth, *Quarterly Journal of Economics*, February, p. 65-94.

Sommer, C. (2001). Der Mittelständler in China - zwischen Mafan und Erfolg, *China-Report*, (34) 15, p. 1-5.

Song, L. (2004). *Beschaffung deutscher Maschinenbauunternehmen in der VR China. Eine praxisorientierte Analyse mit empirischer Untersuchung*, PhD Thesis, Erlangen: Deutscher Universitäts-Verlag.

Spence, M. (1978). Tacit Coordination and Imperfect Information, *Canadian Journal of Economics*, 13, p. 490-505.

Steenhuis, H.-J.; De Bruijn, E. J. (2005). Exploring the Influence of Technology Size on the Duration of Production Technology Transfer Implementation, *International Journal of Technology Transfer and Commercialisation*, 4 (2), p. 172-193.

Steffens, K. (2004). Das WTO-Abkommen über handelsbezogene Aspekte der Rechte des geistigen Eigentums (TRIPS) und das chinesische Recht, in: Heuser, R.; Klein, R.: *Die WTO und das neue Ausländerinvestitions- und Außenhandelsrecht der VR China – Gesetze und Analysen*, Institut für Asienkunde, Hamburg, p. 319-330.

Steinbach, J. (2004). Einfluss des TRIMs-Abkommens auf das chinesische Recht für ausländische Direktinvestitionen, in: Heuser, R.; Klein, R.: *Die WTO und das neue Ausländerinvestitions- und Außenhandelsrecht der VR China – Gesetze und Analysen*, Institut für Asienkunde, Hamburg, p. 67-77.

Stewart, F. (2002). *Arguments for the Generation of Technology by Less-Developed Countries*, Annals of the American Academy of Political and Social Sciences.

Stopford, J.M.; Wells, L.T. (1972). *Managing the Multinational Enterprise*, London.

Swoboda, B. (2002). Dynamische Prozesse der Internationalisierung: Managementtheoretische und empirische Perspektiven des unternehmerischen Wandels, Wiesbaden.

Teece, D. J. (1977). Technology Transfer by Multinational Firms: The Resource Cost of Transferring Technological Know-How, *Economic Journal*, 87 (346), p. 242-261.

Teece, D. J. (2000). *Managing Intellectual Capital: Organizational, Strategic, and Policy Dimensions*, Oxford: Oxford University Press.

Teece, D. J. (1983). Technological and Organizational Factors in the Theory of the Multinational Enterprise, in: Casson, M. (Ed.), *The Growth of International Business*, London, p. 51-62.

Teece, D. J. (1981). The Multinational Enterprise: Market Failure and Market Power Considerations, *Sloan Management Review*, 22 (3), p. 3-17.

Teece, D. J. (1986). Profiting from Technological Innovation: Implications for Integration, Collaboration, Licensing, and Public Policy, *Research Policy*, 15, p. 285-305.

Teich, U.; Lungershausen, D. (1995). Key Points in Articles of Association (Gesell-schaftssatzungen) und Joint-Venture-Verträgen, in: Chung, T. Z.; Sievert, H.-W. (Eds.), *Joint Ventures im chinesischen Kulturkreis*, Wiesbaden.

Terpstra, V.;Yu, C.-M. (1998). Determinants of Foreign Investment of U.S. Advertising Agencies, *Journal of International Business Studies*, 19 (Spring), p. 33-46.

Tesler, L. G. (1980). A Theory of Self-Enforcing Agreements, *Journal of Business*, 53, p. 27-41.

The Technology Atlas Team (1987a). Components of Technology for Resource Transformation, Technological Forecasting and Social Change, 32 (1), p. 19-35.

The Technology Atlas Team (1987b). Assessment of Technology Climate in Two Countries, *Technological Forecasting and Social Change,* 32 (1), p. 85-109.

Thiess, M.; Song, X.; Bernstorf, J. M. (1998). (Bittere) Erfahrungen westlicher Unternehmen in China - Aus Fehlern lernen, in: Karin, D.; Harnischfeger-Ksoll, M. (Eds.), *Erfahrungen im China-Geschäft,* Wiesbaden, p. 4-20.

Thompson, J. D. (1967). *Organizations in Action,* New York: Mac Graw Hill.

Tidd, J.; Izumimoto, Y. (2002). Knowledge Exchange and Learning through International Joint Ventures: An Anglo-Japanese Experience, *Technovation,* 22 (3), p. 137-145.

Toffler, A. (1980). *Die Zukunftschance: Von der Industriegesellschaft zu einer humaneren Zivilisation,* München: Bertelsmann.

Tomlinson, J. W. C. (1970). *The Joint Ventures Process in International Business: India and Pakistan,* Cambridge, MA: The MIT Press.

Trempel, E. J. (2001). China vor dem WTO-Beitritt und die Perspektiven für die Wirtschaft bis zur Olympiade Beijing 2008, in: *Kölner China-Tag,* Deutsche Asia-Pacific-Gesellschaft, Trempel & Associates, Berlin 2001, p. 33.

Ubele, H. (1991). Joint Venture, *Der Betriebswirt,* 51 (2), p. 249-251.

University of Göttingen (2001), *Urheberrechtsgesetz der VR China,* www.jura.uni-goettingen.de/chinarecht/011027.htm.

Vernon, R. (1966). International Investment and International Trade in the Product Cycle, *Quarterly Journal of Economics,* 80 (2), p. 190-207.

Vernon, R. (1974). *Economic Analysis and the Multinational Enterprise.* London: George-Allen & Unwin.

Vernon, R.; Davidson, W. H. (1979). *Foreign Production of Technology-Intensive Products by U.S. Based Multinational Enterprises,* U.S. Department of Commerce.

Vickery, G. (1986). International Flows of Technology – Recent Trends and Developments, *STI Review*, 1, p. 47-83.

Von Behr, M. (2004). Im Sog der Internationalisierung - - Startpunkte, Wege und Ziele kleiner und mittlerer Unternehmen, in: von Behr, M.; Semlinger, K. (Eds.), *Internationalisierung kleiner und mittlerer Unternehmen - Neue Entwicklungen bei Arbeitsorganisation und Wissensmanagement*, München: Campus, p. 45-98.

Von Hippel, E. (1988). *The Sources of Innovation*, Oxford: Oxford University Press.

Von Hippel, E. (1994). Sticky Information and the Locus of Problem Solving: Implications for Innovation, *Management Science*, 40 (4), p. 429-439.

Von Keller, E.; Wei, J.; Drinkuth, H. (2005). *Intellectual Property Protection in China. Playing Weiqi, the Game of Enclosures*. Roland Berger Strategy Consultants.

von Pierer, H. (2004). Speech at the "2. Deutsch-Chinesischen Wirtschafts-kongress", September 2004, Berlin.

von Zedtwitz, M.; Gassmann O. (2002). Market versus Technology Drive in R&D Internationalization: Four Different Patterns of Managing Research and Development, *Research Policy*, 31, p. 569–588.

Weiss, C. A. (1996). *Die Wahl internationaler Markteintrittsstrategien: Eine transaktionskostenorientierte Analyse*, Wiesbaden: Gabler Verlag.

Welge, M. K., Holtbrügge, D. (2003a). *Internationales Management*, Third Edition, Stuttgart.

Welge, M. K.; Holtbrügge, D. (2003b). Kooperative Internationalisierungs-strategien, in: Holtbrügge, D. (Ed.), *Management Multinationaler Unter-nehmungen*, Heidelberg.

Wells, L. T. (1972). Test of a Product Life Cycle Model of International Trade: U.S. Exports of Consumer Durables, in: Wells, L. (Ed.), *The Product Life*

Cycle and International Trade, Boston, MA: Harvard Business School Press, p. 55-79.

Wells, L. T.; Stobaugh, R. B. (1984). *Technology Crossing Borders: The Choice, Transfer, and Management of International Technology Flows.* Harvard Business School Press.

Wernerfelt, B. (1984). A Resource-Based View of the Firm, *Strategic Management Journal*, 5 (2), p. 171-180.

Westney, D. E. (1988). Domestic and Foreign Learning Curves in Managing International Cooperative Strategies, in: Contractor, F.; Lorange, P. (Eds.), *Cooperative Strategies in International Business,* Lexington, MA: Lexington Books, p. 339-346.

Wilkins, A. (1989). *Developing Corporate Character*, San Francisco: Jossey-Bass.

Williamson, O. E. (1975). *Markets and Hierarchies: Analysis and Antitrust Implications. A Study in the Economics of Internal Organization*, New York: The Free Press.

Williamson, O. E. (1979). Transaction-Cost Economics: The Governance of Contractual Relations, *Journal of Law and Economics*, 22, p. 233-261.

Williamson, O. E. (1985). *The Economic Institutions of Capitalism. Firms, Markets, Relational Contracting*, New York: Free Press.

Wilson, R.W. (1977). The Effect of Technological Environment and Product Rivalry on R&D Effort and Licensing of Inventions, *Review of Economics and Statistics*, 59, p. 253-264.

Winship, C.; Morgan, S. L. (1999). The Estimation of Causal Effects From Observational Data, *Annual Review of Sociology*, 25, p. 695-706.

Wolf, B. (1977). Industrial Diversification and Internalization: Some Empirical Evidence, *Journal of Industrial Economics*, 26 (2), p. 177-191.

Woodward, J. (1958). *Management and Technology,* London: H.M.S.O.

Woodward, J. (1965). *Industrial Organization: Theory and Practice*, London: Oxford University Press.

World Bank (1985). *Managing Technological Development, World Bank Staff Working Paper No. 717*, World Bank, Washington D.C.

World Economic Forum (1995). *The World Competitiveness Report 1995*.

World Investment Report 2004, United Nations Conference on Trade and Development.Wrigley, B. A. (2002). IP Value Extraction: Patent and Technology Licenses, *Corporate Counsel*, H., September Issue.

Yan, A.; Gray B. (1994). Bargaining Power, Management Control, and Performance in United States-China Joint Ventures: A Comparative Case Study, *Academy of Management Journal*, 37, p. 1478-1517.

Yan, Y. (2000). *International Joint Ventures in China: Ownership, Control and Performance*, Studies on the Chinese Economy, Basingstoke and London: Macmillan Press Ltd.

Yue, Bing (1997). *Der chinesische Weg zu einer Marktwirtschaft: Erforschung der Strategie der chinesischen Wirtschaftsreform*, PhD Thesis, Gießen University.

Zack, M.H. (1999). Managing Codified Knowledge, *Sloan Management Review*, 40 (4), p. 45-58.

Zaheer, S. (1995). Overcoming the Liability of Foreignness, *Academy of Management Journal*, 38 (2), 341-364.

Zhao, H. (1995). Technology Imports and their Impacts on the Enhancement of China's Indigenous Technological Capability, *Journal of Development Studies*, 31 (4), p. 535-602.

Zielke, A. E. (1992). *Erfolgsfaktoren Internationaler Joint Ventures*, Frankfurt am Main.

Zinzius, B. (2000). *China Business: Der Ratgeber zur erfolgreichen Unternehmensführung im Reich der Mitte*. Berlin: Springer Verlag.

Zürl, K.-H.; Huang, Ji. (2002). *Wirtschaftshandbuch China*, München: Oldenburg.

2. Important Chinese Laws and Regulations

Online links last verified on October 12th, 2007.

Company Law of the People's Republic of China (*Company Law*), last accessed at: http://www.chinalaw.gov.cn/jsp/jalor_en/disptext.jsp?recno=2&&ttlrec=19

Foreign Trade Law of the People's Republic of China (*Foreign Trade Law*), last accessed at: http://www.trade.gov.cn/english/php/show.php?id=9

Law of The People's Republic of China on Wholly Foreign-Owned Enterprises (*WFOE-Law*), last accessed at:
http://en.scnjw.gov.cn/ccm/content/scnjw_english/policy/domestic/
policy13.zh;jsessionid=a9vJzAyGW1j-

Law of the People's Republic of China on Chinese-foreign Contractual Joint Ventures (*CJV-Law*), last accessed at:
http://www.chinalaw.gov.cn/jsp/jalor_en/disptext.jsp? recno=34&&ttlrec=55

Law of the People's Republic of China on Chinese-foreign Equity Joint Ventures (*EJV-Law*), last accessed at:
http://www.chinalaw.gov.cn/jsp/jalor_en/disptext.jsp?recno=67&&ttlrec=111

Provisions of the State Council of the People's Republic of China for the Encouragement of Foreign Investment, last accessed at:
http://www.novexcn.com/encour_foregn_invest.html

Provisions of the State Council of the People's Republic of China on Guiding Direction of Foreign Investment, last accessed at
http://www.gov.cn/english/laws/2005-07/25/content_16873.htm

Regulations for the Implementation of the Law of the People's Republic of China on Chinese-foreign Equity Joint Ventures, last accessed at:
http://www.chinalaw.gov.cn/jsp/jalor_en/ disptext.jsp?recno=59&&ttlrec=111

Regulations on Technology Import and Export Administration of the People's Republic of China, last accessed at: http://www.ccpit-
patent.com.cn/references/Regulations_Technology
_Import_Export_Administration_China.htm#ChapterI

The Catalogue for the Guidance of Foreign Investment Industries, <u>last accessed</u> <u>at: http://www1.cei.gov.cn/ce/index/report/cep7/guid2004.htm#part2</u> on October 11th, 2007.

3. Appendices

Appendix 1: Technology Fields

Code	Technology Fields
BA	Mining, civil engineering, airconditioning, building materials, waste disposal
PP	Paper, printing
TE	Textiles, apparel, leisure, textile machinery
ME	Biomedical engineering (biomedicine)
NA	Agriculture, nutrition, beverages, tobacco
GP	Bio- and genetic engineering (genetics), pharmacy
OC	Organic chemistry, petrochemistry
PC	Polymer materials (polymer chemicals)
SY	Manufacturing & application of polymers (synthetic resins, paints, etc.)
IC	Inorganic chemistry, glass, explosives
CO	Coating, crystal growing
SM	Process engineering, separation, mixing
MA	Mechanical engineering, machinery, armament
MM	Material processing, machine tools
HA	Handling, conveyor equipment, robots
TR	Transport, traffic
ET	Engines, turbines, pumps
EN	Electric power, nuclear technology
EM	Electrical machinery
LA	Lasers
OP	Optical equipment
IN	Instruments, controls
MS	Metrology, sensors
DA	Data processing
IS	Information storage
TC	Telecommunication (not image transmission)
IM	Image transmission
EL	Electronics, electronic components

Source: adapted from Engelsmann and van Raan (1994).

261

Appendix 2: Levels of Sophistication of Technological Resources

	Level of Sophistication	Examples
Technoware		
1.	Manual tools	Screwdriver, hand drill
2.	Powered equipment	Grinder, power drill
3.	General purpose facilities	Milling machine, lathe
4.	Special purpose facilities	Textile power looms, airjet weaving loom
5.	Automatic machines	Soft drink bottling plant
6.	Computerised facilities	Numerical Control (NC) machines
7.	Integrated facilities	Completely robotised assembly plants, integrated plants
Inforware		
1.	Familiarising facts	Brochure, images
2.	Describing facts	Technical booklet, process description
3.	Specifying facts	Performance and usage specifications
4.	Utilising facts	Standard operating and maintenance manuals
5.	Comprehending facts	Process theories, design data and calculations
6.	Generalising facts	Development information generated through indigenous R&D
7.	Assessing facts	Comprehensive information on the latest developments
Humanware		
1.	Operating abilities	Unskilled and semi-skilled operators
2.	Setting-up abilities	General technicians, skilled operators
3.	Repairing abilities	Special technicians, maintenance engineers
4.	Reproducing abilities	Production engineers
5.	Adapting abilities	Design engineers
6.	Improving abilities	Development engineers (development)
7.	Innovating abilities	Development engineers (research)
Orgaware		
1.	Individual linkages	Small firm
2.	Collective linkages	Connected small firms
3.	Departmental linkages	Small-scale organisation
4.	Enterprise linkages	Medium-scale organisation
5.	Industrial linkages	Large-scale organisation
6.	National linkages	Multi-location organisation
7.	Global linkages	Transnational organisation

Source: adapted from Cohen (2004), the Technology Atlas Team (1987a) and M.Sharif (1988).

Appendix 3: Description of Expert Interviews

No	Description of the Organisation	Role of the Interviewee	General Description of Interview
1	GermanCenter, Beijing	Assistant to the director	The GermanCenter is a 100% subsidiary of the Landesbank Baden-Württemberg. The institution consults German firms, especially SMEs, on the topic of market entry to China.
2	An independent, medium-sized steel maker supplying various industries.	Chief representative in China.	The firm has many years of experience in cooperations with China, including projects for large state-owned enterprises or the government.
3	Tsinghua university	Professor of European origin teaching in the university	The professor has many years of experience in consulting firms on how to operate in the Chinese market and runs his own consulting firm.
4	Major technology firm based in Europe that operates in a variety of industries.	Head of R&D China	The firm has many years of experience in China and currently is involved in major projects in the energy sector. The Chinese operations contribute to the global R&D.
5	One of the largest automobile manufacturers in Europe	Former senior consultant who is now Professor in China	The company has many years of experience of operating in China and is involved in several joint ventures for many years.
6	A manufacturer of brake systems and vibration dampers for rail and commercial vehicles.	Assistant who did the technical documentation for a JV establishment	The contact person was involved in preparing the technical documentation for one of the JVs that the firm has invested in in China.
7	Large chemical company that produces additives for e.g. the automotive, the personal care and the plastics industry.	Head of corporate development in China	The firm maintains 10-15 factories in China. Since 2005, it has the motto 'invented in China' and each of the three business divisions conducts local development.
8	Global European enterprise with core competencies in the fields of health care, nutrition and high-tech materials	Plant manager of integrated production site in China & head of technology services group	The firm has been present in China for more than 100 years. All corporate divisions are present with manufacturing sites. The firm is invested in WFOEs as well as joint ventures.
9	International law firm with offices in China	Attorney in the Chinese office	The attorney consults foreign investors on the legal implementation of their market entry strategy. The respective expert has accompanied several joint venture establishments before.

Appendix 4: Overview of Variables that Correspond to Hypotheses

	Variable Name	Questionnaire Questions (all scales from 1= " don't agree" to 5 = "I agree" unless stated otherwise*)
Hypothesis 1a - d		
1a) Costs of own cooperation	C_{AC}	V050c: The costs of sharing our own knowledge with the partner and thus making it less "unique" are high
		V050d: The costs of the potential diffusion of our own knowledge among the industry are high
1b) Costs of own defection	C_{AD}	V051b: Uncooperative behaviour by us would lead to high costs for us due to image loss
		V051c: Uncooperative behaviour by us could lead to long-term damage to our company
1c) Costs of being defected upon	C_{BD}	V051d: Uncooperative behaviour by the partner would lead to high damage to our company
1d) Benefits from mutual cooperation (Synergy)	V_S	V050a: There is high complementarity between the resources/capabilities of the two partners
		V050b: Via collaboration and knowledge exchange, we are able to create strong synergies
Hypothesis 2		
2) Trust	Trust	V048c: The JV is characterised by mutual trust between the partners at multiple levels
Hypothesis 3a & b		
3a) Age of JV	AgeJV	V015a: When was the Joint Venture established (month, year)? - Translated into age in months
3b) Duration of Contract	LengthContract	V015b: How long is the JV contract for (in years)?
Hypothesis 4a & b		
4a) Discount Rate	DiscountRate	V009: The discount rate that is used for the evaluation of foreign investments like Joint Ventures in China in %
4b) Time Horizon	TimeHorizon	V008: Time horizon applied for the evaluation of foreign investments in years
Hypothesis 5a & b		
5a) Frequency of Interaction	FreqInteract	V048a: There is close, personal interaction between the partners at multiple levels
		V049e: At an operative level, communication between partners takes place frequently
5b) Behavioural Transparency	Transparency	V049c: After how much time does your firm typically learn about changes in your partner's behaviour (such as compliance or non-compliance with the agreement) related to your firm's alliance? (scale from 1 to 5: 1= within 1 day, 5 = over one month)
		V049d: We usually have accurate information about the partner's behaviour
Hypothesis 6a & b		
6a) Multiple Links	Links	V049j: There exist links between our two firms due to several projects, not only due to the JV
6b) Dependency	Dependency	V052b: After the JV contract ends, there will still be mutual dependency between us both

*where one variable name covers several questionnaire questions, the variable was computed by calculating the mean of the respective questions.

264

Appendix 5: Overview of General Influence Factors and Corresponding Variables

Concept	Variable Name	Questionnaire Questions (all scales from 1=" don't agree" to 5 ="I agree" unless stated otherwise*)
Local Market Conditions		
Local market conditions.	LocalCond	V029: The country-specific requirements in China make an adaptation of our products necessary
Customer-specific requirements	CustomReq	V030a-d: Our customers in China are very demanding with respect to: a) innovative products b) safety of products c) short response time d) good service
		V031: The bargaining power of our customers is high.
Compliance of output with regulations	OutputCompl	V033: Compliance with local product regulations makes the adaptation of our products unavoidable
IP Regime needed to loss risk	IPLossRisk	V036: We need to rely on a strong IP regime because the risk of knowledge diffusion and imitation by competitors is high
Industry-Specific Market Attractiveness		
Market attractiveness	MarketAttract	V028a: How did you assess the product's market size in China?
		V028b: How did you assess the product's market potential in China?
Differences in Cost Levels		
Differences in Cost Levels	CostLevels	V035a-d: Has the cost advantage of a) raw materials b) low-skilled workers c) high-skilled workers d) infrastructure promoted the investment in local activities?
Government Incentives		
Tender	Tender	V032a: The JV is a result of a tender by the Chinese government or a public enterprise, in which technology transfer was one component for winning the tender. (1=yes, 0=no)
Incentives	GovIncentives	V032b: Government incentives for investing (technology) in the JV were (1= low, 5= high)
Entry Regulations	Preference.JV	V018a: Was the choice "Joint Venture" your preferred market entry form in this case? (1= yes, 0=no)
Characteristics of the Technology Sending Firm		
Firm Size	FirmSize	V002: How many employees did the company have at the end of 2006?
Degree of internationalisation (TNI)	FirmTNI	V004a-c: a) Foreign Assets/ Total Assets in % (2006), b) Foreign Sales/ Total Sales in % (2006), c) Foreign Employees/ Total Employees in % (2006)
Experience in China	ExpChina	V007a-c: In China, number of a) Joint Ventures established b) accomplished international technology transfer projects c) all foreign direct investments
	GlobalIntegr	V005a: The company's commitment to global integration of activities is high
	LocalAdapt	V005b: The company's commitment to local adaptation of activities is high
	OnwershipShare	V012c: The investor's ownership share in %
Firm Strategy & JV Role	EquityJV	V012a: What type of Joint Venture is it? (Contractual JV = 0, Equity Joint Venture = 1)
	JVType	V013: Technology-to-market JV: "1", Sales JV: "2", Complementary technology JV: "3", Concentration JV: "4", Sourcing JV: "5", R&D JV: "6", Production JV: "7"
Competitive Pressure	CompPressure	V045: At the time of the JV's establishment, we were under high competitive pressure regarding market entry to China

(continued)

Concept	Variable Name	Questionnaire Questions (all scales from 1= " don't agree" to 5 = "I agree" unless stated otherwise")
Product and Technology Characteristics		
Resource Costs of transfer	LowResourceCosts	V040a: The described technology that has been transferred by us to the JV was used by us for the production of that product x years ago for the first time.
		V040b: The described technology that has been transferred by us to the JV has been transferred x times to foreign plants
		V040e: The described technology that has been transferred by us to the JV is well-known and widespread in our industry
Inherent Replicability	Replicability	V041: The technology that has been transferred by us is a) very complex b) easy to document c) easy to teach
Strategic Value to the Firm	TTCore	V040c: The described technology that has been transferred by us to the JV is a strategically important technology (core competence).
	TTModern	V040d: The described technology that has been transferred by us to the JV is modern when compared to our most modern plants in the
The Technology Recipient and Practical Implementation Barriers		
Absorptive Capacity	Absorptive	V043a: The Joint Venture partner has a good understanding of technologies
		V043b:The Joint Venture partner has a high absorptive capacity when learning new technologies
Transfer Barriers	Barriers	V044a: As what do you assess barriers to technology transfer due to cultural barriers to be for the given case?
		V044b: As what do you assess barriers to technology transfer due to communication barriers to be for the given case?
		V044c: As what do you assess barriers to technology transfer due to goal divergence to be for the given case?
Bargaining Power	BargInvestor	V053a: When negotiating the JV contract, our management had a higher bargaining power than the partner did

"where one variable name covers several questionnaire questions, the variable was computed by calculating the mean of the respective questions. An exception is the variable LowResourceCosts. Here, the questions V040a and V040b were transformed into a scale from 1-5 before calculating the average. If LowResourceCosts is equal to 5, the resource costs for the transfer of technology are comparably low compared to other firms in the sample and vice versa.

Appendix 6: Qualitative Descriptions of Joint Ventures

No.	Described Product	General Description of JV-Setup	Preferred MEF	Described Role of Respondent in JV	Described Role of Chinese Partner in JV
A001	Process automation	Adaptation of imported products and installation and launch of systems.	JV	Complete responsibility and leadership	Investor with expectation for dividends
A002	Surface acoustic wave filters	Manual production with local sourcing.	JV	Capital, technology (product design), export	Capital, technology for manual processes, local know-how
A003	Air separation plant	Almost world-class technology.	JV	Engineering	Sales and facility maintenance
A004	Energy chains for steelplate processing	Autarkic operations in China.	JV	Know-how	Production
A005	Construction paint	Paint production, medium level for EU standards, high level for Chinese standards.	JV	Capital, marketing, product development, controlling	Operative and strategic management of organisation
A006	Laboratory scales	Assembly plant with local sourcing. Start of own development.	JV	Operator of the organisation	Industrial park
A007	Pharmaceutical product	Autarkic production plant.	WFOE	Production and sales	Utilities (water, energy)
A008	Carbon black	Previous JV partner is bankrupt. Current JV partner is not related to the industry.	JV	Know-how, operation competence, brand, everything	Capital, government affairs
A009	Gas pressure springs	Assembly plant. Import of critical parts from Germany. Local sales.	JV	Technology, machines and facilities, products, customers	Capital
A010	Greenhouses	JV is legal construction for China activities. JV partner has virtually never seen the plant.	WFOE	Everything	Network, government affairs
A011	Spark plugs	90% own production. Rest: CKD.	WFOE		
A012	Central lubrication (steel industry)	Completely knocked down (CKD).	JV	Core components, know-how	Local production and sales for China
A013	Brake wheel cylinders		JV		
A014	Packaging	Partner has no own factories. He just added capital and worked as door opener.	WFOE	Money, management, international customers	Connections during JV launch. Today not needed anymore
A015	ATM-DSLAM	Completely knocked down (CKD).	JV	Technology, product, customers	Low cost infrastructure, personnel and customers.
A016	Pharmaceutical product	Simple pharmaceutical production and sales for local market.	WFOE	Everything	No active role
A017	Zinc phosphating	Chemical plant for commodities. World-class technology.	JV	Technology	Market entry

No.	Described Product	General Description of JV-Setup	Preferred MEF	Described Role of Respondent in JV	Described Role of Chinese Partner in JV
A018	Construction paint	Chemical plant for commodities. Simple production facilities with low productivity.	JV	Product know-how, capital	Market know-how, capital
A019	Engine maintenance	Overhaul and repair of air plane engines.	JV	Technology	Capacity utilisation
A020	Passenger car	Full-scale plant.	WFOE	Technology, know-how	Market know-how
A021		Transfer only successful initially, most of it lost by now.	JV	Know-how	Contacts, personnel
A022	Big bags	Operation of imported machines with local workers and inputs.	JV	Know-how, sales	Resources (personnel, land), accounting
A023	Exhaust manifolds	Iron foundry for automobile industry with world-class facilities and knowledge transfer.	JV	Management, technology, investment capital	Relations to important customers, infrastructure
A024	Gaskets	Partially independent organization with medium technology and no competence for development.	JV	Technology, capital, projects, customers	Relatively little
A025	Tailored blanks	World-class technology.	JV	Production facilities, personnel	Infastructure and halls
A026	Screw-type compressors	Assembly. Import of high-tech parts. Local production/sourcing of low-tech parts.	WFOE	Product & process technology, organisation, accounting	Personnel, infrastructure, government affairs
A027	Devices for air separation plants	Except for a few know-how parts equivalent to the production in Germany.	JV	Everything	Partner little involved in operations
A028	Inhalers	High-quality plastic parts for medical and technical applications. World-class technology.	JV	Everything, especially R&D and Sales	Personnel, public relations, government affairs
A029	Gas storage	JV is the bridge to enter China market.	WFOE	Management	Local authority relationship
A030	Truck lamps	Imported parts, local parts, production and assembly, stamp parts, simple technology.	JV	Everything	Too many unskilled workers who are still on the payroll.
A031	Car locking systems	Assembly plant with local development, technical support for local suppliers.	WFOE	Product, technology	Market access (investment regulation)
A032	Thermometers	Important production facility for world supply.	WFOE		
A033	Car locking systems	Assembly plant. Import of critical parts from Germany.	Licensing	Technology, processes, products	Buildings, personnel, contacts
A034	Multi-function switches	Plant on world standard equivalent to Germany.	WFOE	Market and costs	Earnings

* the rows indicated in grey are those where the sophistication level of the partner exceeds the sophistication level of the German firm worldwide or the joint venture in any of the dimensions Technoware, Inforware, Humanware, Orgaware, or Capabilities (useful for discussion in Section IV-2).

Appendix 7: Survey Questionnaire (English Version)

(see next page)

INTERNATIONALES MANAGEMENT
FACULTY OF BUSINESS AND ECONOMICS

SURVEY

Your Joint Venture(s) in China: How much technology have you transferred, and why?

- Any information provided will be treated as confidential and made anonymous -

Thank you very much in advance for participating in our survey!

Please send your completed questionnaire to the following address:

via email to:	michael.hoeck@im.rwth-aachen.de or
via fax to:	+49 (0) 0241-80-92348 or
via letter mail to:	Lehrstuhl für Internationales Management, RWTH Aachen
	Templergraben 64, 6.Stock
	D- 52056 Aachen, Germany

INTERNATIONALES MANAGEMENT
FACULTY OF BUSINESS AND ECONOMICS

1. GENERAL INFORMATION ON THE COMPANY

To start off, please provide some information on the company or the company division in whose name you are completing the questionnaire.

1. For what company are you responding to **this questionnaire**?

2. How many **employees** did the company have at the end of 2006?

3. What was the total amount of **sales** generated in 2006?
   (in € million)

4. How **international** is your company according to the following indicators (2006 figures)?

Share of Foreign Assets	/ Total Assets	in %
Share of Foreign Sales	/ Total Sales	in %
Share of Foreign Employees	/ Total Employees	in %

5. Positioning of the company with regard to global integration / local adaptation of activities:

   a) The company's commitment to **global Integration** of activities is high.
   I don't agree   1. ☐   2. ☐   3. ☐   4. ☐   5. ☐   I agree
   b) The company's commitment to **local adaptation** of activities is high.
   I don't agree   1. ☐   2. ☐   3. ☐   4. ☐   5. ☐   I agree

6. The company has **world-wide** experience regarding the following activities:

   Number of Joint Ventures established:
   Number of accomplished international technology transfer projects:
   Number of all foreign direct investments:

7. The company has experience regarding the following activities in **China**:

   Number of Joint Ventures established:
   Number of accomplished international technology transfer projects:
   Number of all foreign direct investments:

8. The **time horizon** applied for the evaluation of foreign investments is about:
   years

9. If you use the „Discounted Cash Flow" (DFC)-method to evaluate investments: The **discount rate** that is used for the evaluation of foreign investments like Joint Ventures in China is:
   %

2. GENERAL INFORMATION ON THE JOINT VENTURE ('JV')

*Please answer the following questions regarding the Joint Venture. If your company has invested in several Joint Ventures in China, please answer with respect to **one specific** Joint Venture.*

10. What is the **Joint Venture** called? (answer is optional)

11. Who is the **Chinese partner**? (answer is optional)

12. What **type** of Joint Venture is it? ☐ Contractual JV   ☐ Equity JV

If the JV is an Equity Joint Venture: what is the total registered capital and your ownership share?

Registered capital in €   Your ownership share in %

13. The Joint Venture can best be described as a:
   ☐ Technology-to-market JV   ☐ Sales JV
   ☐ Complementary technology JV   ☐ Concentration JV
   ☐ Sourcing JV   ☐ R&D JV
   ☐ Production JV   Other:

14. Are there more than two partners involved? If so, what is the total number of partners?

If there are more than 2 partners involved, please refer to the Chinese partner with the highest share in the JV when answering all following questions.

15. Please provide some **basic information** regarding the Joint Venture:
   a) When was the Joint Venture established (month, year)?
   b) How long is the JV contract for?   years
   c) How many employees were there in total at the end of 2006?
   d) How high was the sales revenue in 2006?   €
   e) How high was the export share in 2006 in %   %

16. Which **functions** are covered by the Joint Venture?
   ☐ Research   ☐ Development   ☐ Purchasing   ☐ Production
   ☐ Marketing   ☐ Sales   ☐ Service   ☐ Accounting

17. How would you assess the **JV's success** in 2006 in comparison to the company-wide average?

	Low				High
Return on investment	1. ☐	2. ☐	3. ☐	4. ☐	5. ☐
Profit	1. ☐	2. ☐	3. ☐	4. ☐	5. ☐
Market share	1. ☐	2. ☐	3. ☐	4. ☐	5. ☐

18. Was the choice "Joint Venture" your **preferred market entry form** in this case?
   Yes, a JV was the preferred market entry form.   ☐
   No, we would have preferred:

19. All of the following questions should be answered in reference to **one** of the **products, product groups or services** that are produced by the JV. Please describe the chosen product and state its share of the JV's turnover in 2006:

Product: Share of turnover in %:

20. How would you characterize the **production process** for that product?
   ☐ Unit or small batch production   ☐ Large batch or mass production
   ☐ Process production   ☐ Combined systems

3. TECHNOLOGICAL ENDOWMENT OF THE JOINT VENTURE

*In the following section, we will ask you for information on the technology employed in the Joint Venture. Please refer to the above **chosen product (group)** when providing the information.*

21. Please describe up to three of the **most important technologies** or skills that are necessary for the production of the described product (e.g."float glass manufacturing" or "fermentation").

Technology 1:
Technology 2:
Technology 3:

22. The **technology** that is employed by the JV is
a) innovative.

I don't agree	1. ☐	2. ☐	3. ☐	4. ☐	5. ☐	I agree

b) complex.

I don't agree	1. ☐	2. ☐	3. ☐	4. ☐	5. ☐	I agree

23. Please indicate which **maximum technological level** with respect to the chosen product is being achieved by your company world-wide (**C**), by the Joint Venture (**JV**), and by the Chinese partner (**P**):
(i.e. three ticks/checks per dimension)

Levels	Examples	C	JV	P
Dimension 1: "Technoware"				
1. Manual tools	Screwdriver, hand drill	☐	☐	☐
2. Powered equipment	Grinder, power drill	☐	☐	☐
3. General purpose facilities	Milling machine, lathe	☐	☐	☐
4. Special purpose facilities	Textile power looms, airjet weaving loom	☐	☐	☐
5. Automatic machines	Soft-drink bottling plant	☐	☐	☐
6. Computerized facilities	Numerical Control (NC) machines	☐	☐	☐
7. Integrated facilities	Completely robotized assembly plants, integrated plants	☐	☐	☐
Dimension 2: "Inforware"				
1. Familiarizing facts	Brochure, images	☐	☐	☐
2. Describing facts	Technical booklet, process description	☐	☐	☐
3. Specifying facts	Performance and usage specifications	☐	☐	☐
4. Utilizing facts	Standard operating and maintenance manuals	☐	☐	☐
5. Comprehending facts	Process theories, design data and calculations	☐	☐	☐
6. Generalizing facts	Development information generated through indigenous R&D	☐	☐	☐
7. Assessing facts	Comprehensive information on the latest developments	☐	☐	☐
Dimension 3: "Humanware"				
1. Operating abilities	Unskilled and semi-skilled operators	☐	☐	☐
2. Setting-up abilities	General technicians, skilled operators	☐	☐	☐
3. Repairing abilities	Special technicians, maintenance engineers	☐	☐	☐
4. Reproducing abilities	Production engineers	☐	☐	☐
5. Adapting abilities	Design engineers	☐	☐	☐
6. Improving abilities	Development engineers (development)	☐	☐	☐
7. Innovating abilities	Development engineers (research)	☐	☐	☐
Dimension 4: "Orgaware"				
1. Individual linkages	Small firm	☐	☐	☐
2. Collective linkages	Connected small firms	☐	☐	☐
3. Departmental linkages	Small-scale organization	☐	☐	☐
4. Enterprise linkages	Medium-scale organization	☐	☐	☐
5. Industrial linkages	Large-scale organization	☐	☐	☐
6. National linkages	Multi-location organization	☐	☐	☐
7. Global linkages	Transnational organization	☐	☐	☐

24. Please indicate your assessment of the **technological capabilities** of your company (headquarters), of the Joint Venture, and of your Chinese partner (1=weak, 5=strong):

a) Operative capabilities
Operating and controlling plant and equipment, planning and controlling production activities, providing information support and networking for operations, keeping the plant and equipment in good working order.

Company	1. ☐	2. ☐	3. ☐	4. ☐	5. ☐
Joint Venture	1. ☐	2. ☐	3. ☐	4. ☐	5. ☐
Partner	1. ☐	2. ☐	3. ☐	4. ☐	5. ☐

b) Acquisitive capabilities
Carrying out a detailed engineering study, independently searching for good technology sources, assessing technologies offered, deciding on technology transfer mode, and negotiating terms of technology transfer.

Company	1. ☐	2. ☐	3. ☐	4. ☐	5. ☐
Joint Venture	1. ☐	2. ☐	3. ☐	4. ☐	5. ☐
Partner	1. ☐	2. ☐	3. ☐	4. ☐	5. ☐

c) Innovative capabilities
Duplicating acquired technology, adopting and carrying out improvements in imported technology, and carrying out own technology development plan.

Company	1. ☐	2. ☐	3. ☐	4. ☐	5. ☐
Joint Venture	1. ☐	2. ☐	3. ☐	4. ☐	5. ☐
Partner	1. ☐	2. ☐	3. ☐	4. ☐	5. ☐

d) Supportive capabilities
Project planning, project financing, planning and implementing human resource development, and identifying and developing new markets for the firm's existing and new products.

Company	1. ☐	2. ☐	3. ☐	4. ☐	5. ☐
Joint Venture	1. ☐	2. ☐	3. ☐	4. ☐	5. ☐
Partner	1. ☐	2. ☐	3. ☐	4. ☐	5. ☐

25. How would you characterize the **Chinese partner's** degree of insight?

a) The Chinese partner understands all of the technologies employed by the JV.

I don't agree 1. ☐ 2. ☐ 3. ☐ 4. ☐ 5. ☐ I agree

b) The Chinese partner could produce the JV's products by himself.

I don't agree 1. ☐ 2. ☐ 3. ☐ 4. ☐ 5. ☐ I agree

c) The Chinese partner's knowledge surpasses our own in relevant technological areas.

I don't agree 1. ☐ 2. ☐ 3. ☐ 4. ☐ 5. ☐ I agree

26. Value added share of the end product:
When referring to the **weight proportions**, the Joint Venture locally produces _____ % of the end product.
When referring to **value proportions**, this would be around _____ %.

27. How would you describe the **role of the JV** and its technological endowment (possibly also with implications for knowledge transfer to the Chinese partner)?

Examples:
- "Completely knocked down" production site, strategic parts are imported *or*
- Chemical plant for commodities with world-class technology, producing on a world scale

4. EXTERNAL ENVIRONMENT

*The following section will ask you for information regarding the external environment that prevailed during **the establishment phase of the Joint Venture**. Again, please relate your answers to the chosen product.*

28. What did you assess the **Chinese market's** attractiveness for the product to be?

	Low				High
Market size	1. ☐	2. ☐	3. ☐	4. ☐	5. ☐
Market potential	1. ☐	2. ☐	3. ☐	4. ☐	5. ☐

29. The country-specific **requirements** (socio-cultural, economic, natural) in China make an adaptation of our products necessary.

I don't agree   1. ☐   2. ☐   3. ☐   4. ☐   5. ☐   I agree

30. Our **customers** in China are very demanding with respect to:

	I don't agree				I agree
Innnovative products	1. ☐	2. ☐	3. ☐	4. ☐	5. ☐
Safety of products	1. ☐	2. ☐	3. ☐	4. ☐	5. ☐
Short response time	1. ☐	2. ☐	3. ☐	4. ☐	5. ☐
Good service	1. ☐	2. ☐	3. ☐	4. ☐	5. ☐

31. The **bargaining power** of our customers is high.

I don't agree   1. ☐   2. ☐   3. ☐   4. ☐   5. ☐   I agree

32. a) The JV is a result of a **tender** by the Chinese government or a public enterprise, in which technology transfer was one component for winning the tender.
☐ Yes ☐ No

b) Government **incentives** for investing (technology) in the JV were

Low   1. ☐   2. ☐   3. ☐   4. ☐   5. ☐   High

33. Compliance with local product regulations makes the adaptation of our products unavoidable.

I don't agree   1. ☐   2. ☐   3. ☐   4. ☐   5. ☐   I agree

34. Has the **quality advantage** of inputs promoted the investment in local activities?

	I don't agree				I agree
Raw materials	1. ☐	2. ☐	3. ☐	4. ☐	5. ☐
Low-skilled workers	1. ☐	2. ☐	3. ☐	4. ☐	5. ☐
High-skilled workers	1. ☐	2. ☐	3. ☐	4. ☐	5. ☐
Infrastructure	1. ☐	2. ☐	3. ☐	4. ☐	5. ☐

35. Has the **cost advantage** of inputs promoted the investment in local activities?

	I don't agree				I agree
Raw materials	1. ☐	2. ☐	3. ☐	4. ☐	5. ☐
Low-skilled workers	1. ☐	2. ☐	3. ☐	4. ☐	5. ☐
High-skilled workers	1. ☐	2. ☐	3. ☐	4. ☐	5. ☐
Infrastructure	1. ☐	2. ☐	3. ☐	4. ☐	5. ☐

36. Companies in our industry have to rely on a strong **intellectual property** right regime because the risk of knowledge diffusion and imitation by competitors is high.

I don't agree   1. ☐   2. ☐   3. ☐   4. ☐   5. ☐   I agree

37. According to the **catalogue** of the Chinese government, our industry belongs to the .... industries.
☐ promoted   ☐ restricted   ☐ not specially treated

INTERNATIONALES MANAGEMENT
FACULTY OF BUSINESS AND ECONOMICS

5. PRODUCT, PROJECT AND TECHNOLOGY SPECIFIC FACTORS

In the following section, we will ask you for information on the product and the technology that you have transferred to the JV. Please refer to the chosen product again.

38. The product is at the following stage of its **life cycle**:

	Introduction	Growth	Maturity	Saturation	Decline
In Europe	1. ☐	2. ☐	3. ☐	4. ☐	5. ☐
In China	1. ☐	2. ☐	3. ☐	4. ☐	5. ☐

39. The **product** that is produced in China is
a) by world standards advanced.

I don't agree	1. ☐	2. ☐	3. ☐	4. ☐	5. ☐	I agree

b) strategically important for us.

I don't agree	1. ☐	2. ☐	3. ☐	4. ☐	5. ☐	I agree

c) mainly designated to be sold in the Chinese market.

I don't agree	1. ☐	2. ☐	3. ☐	4. ☐	5. ☐	I agree

40. The described technology that has been **transferred by us** to the JV
a) was used by us - for the production of that product -   years ago for the first time.
b) has been transferred   times to foreign plants.
c) is a strategically important technology (core competence).

I don't agree	1. ☐	2. ☐	3. ☐	4. ☐	5. ☐	I agree

d) is modern when compared to our most modern plants in the world.

Out-dated	1. ☐	2. ☐	3. ☐	4. ☐	5. ☐	Equally modern

e) is well-known and widespread in our industry.

I don't agree	1. ☐	2. ☐	3. ☐	4. ☐	5. ☐	I agree

41. The technology that has been **transferred by us** is

	I don't agree				I agree
very complex.	1. ☐	2. ☐	3. ☐	4. ☐	5. ☐
easy to document.	1. ☐	2. ☐	3. ☐	4. ☐	5. ☐
easy to teach.	1. ☐	2. ☐	3. ☐	4. ☐	5. ☐

42. Please estimate what percentage of the **total project costs** for installing the product line in the JV can be assigned to the following cost factors of the technology transfer:

	0 - 2%	2 - 4 %	4 – 6 %	6 – 8%	> 8%
Pre-engineering technological exchanges	1. ☐	2. ☐	3. ☐	4. ☐	5. ☐
Engineering costs for process design	1. ☐	2. ☐	3. ☐	4. ☐	5. ☐
R&D personnel during all phases of project	1. ☐	2. ☐	3. ☐	4. ☐	5. ☐
Start-up training and "excess-manufacturing"	1. ☐	2. ☐	3. ☐	4. ☐	5. ☐

43. The Joint Venture **partner**
a) has a good understanding of technologies.

I don't agree	1. ☐	2. ☐	3. ☐	4. ☐	5. ☐	I agree

b) has a high absorptive capacity when learning new technologies.

I don't agree	1. ☐	2. ☐	3. ☐	4. ☐	5. ☐	I agree

44. As what do you assess **barriers** to technology transfer to be for the given case?

	Low				High
Cultural barriers	1. ☐	2. ☐	3. ☐	4. ☐	5. ☐
Communication barriers	1. ☐	2. ☐	3. ☐	4. ☐	5. ☐
Goal divergence	1. ☐	2. ☐	3. ☐	4. ☐	5. ☐

45. At the time of the JV's establishment, we were under high **competitive pressure** regarding market entry to China.

I don't agree	1. ☐	2. ☐	3. ☐	4. ☐	5. ☐	I agree

INTERNATIONALES MANAGEMENT
FACULTY OF BUSINESS AND ECONOMICS

6. RELATIONSHIP WITH JV PARTNER

Please provide some answers regarding your relationship with the JV partner.

46. Please indicate the **relative responsibilites** within the JV:

	Only us		both		Only partner	n.a.
Research	1. ☐	2. ☐	3. ☐	4. ☐	5. ☐	☐
Development	1. ☐	2. ☐	3. ☐	4. ☐	5. ☐	☐
Purchasing	1. ☐	2. ☐	3. ☐	4. ☐	5. ☐	☐
Production	1. ☐	2. ☐	3. ☐	4. ☐	5. ☐	☐
Accounting	1. ☐	2. ☐	3. ☐	4. ☐	5. ☐	☐
Sales	1. ☐	2. ☐	3. ☐	4. ☐	5. ☐	☐
Service	1. ☐	2. ☐	3. ☐	4. ☐	5. ☐	☐
Marketing	1. ☐	2. ☐	3. ☐	4. ☐	5. ☐	☐

47. What do your company and your partner company mainly **contribute** to the JV?
Your company:
Your JV partner:

48. Please assess the following claims regarding the **history of interaction** between you and your partner:

a) There is close, personal interaction between the partners at multiple levels.
I don't agree   1. ☐   2. ☐   3. ☐   4. ☐   5. ☐   I agree
b) The JV is characterized by mutual respect between the partners at multiple levels.
I don't agree   1. ☐   2. ☐   3. ☐   4. ☐   5. ☐   I agree
c) The JV is characterized by mutual trust between the partnes at multiple levels.
I don't agree   1. ☐   2. ☐   3. ☐   4. ☐   5. ☐   I agree
d) The JV is characterized by personal friendship between partners at multiple levels.
I don't agree   1. ☐   2. ☐   3. ☐   4. ☐   5. ☐   I agree

49. Please provide information regarding the **current interaction** between you and your partner:

a) The JV is characterized by high reciprocity among the partners.
I don't agree   1. ☐   2. ☐   3. ☐   4. ☐   5. ☐   I agree
b) There exist links between our two firms due to several projects, not only due to the JV.
I don't agree   1. ☐   2. ☐   3. ☐   4. ☐   5. ☐   I agree
c) After how much time does your firm typically learn about changes in your partner's behavior (such as compliance or non-compliance with the agreement) related to your firm's alliance?
Within 1 day   1. ☐   2. ☐   3. ☐   4. ☐   5. ☐   > one month
d) We usually have accurate information about the partner's behaviour.
I don't agree   1. ☐   2. ☐   3. ☐   4. ☐   5. ☐   I agree
e) At an operative level, communication between partners takes place frequently.
I don't agree   1. ☐   2. ☐   3. ☐   4. ☐   5. ☐   I agree

50. Please assess the following claims regarding the **benefits and costs of knowledge sharing**:

a) There is high complementarity between the resources/capabilities of the two partners.
I don't agree   1. ☐   2. ☐   3. ☐   4. ☐   5. ☐   I agree
b) Via collaboration and knowledge exchange, we are able to create strong synergies.
I don't agree   1. ☐   2. ☐   3. ☐   4. ☐   5. ☐   I agree
c) The costs of sharing our own knowledge with the partner and thus making it less "unique" are high.
I don't agree   1. ☐   2. ☐   3. ☐   4. ☐   5. ☐   I agree
d) The costs of the potential diffusion of our own knowledge among the industry are high.
I don't agree   1. ☐   2. ☐   3. ☐   4. ☐   5. ☐   I agree

INTERNATIONALES MANAGEMENT
FACULTY OF BUSINESS AND ECONOMICS

51. Please assess the following claims regarding the **costs** of (potential) **uncooperative behaviour**:

a) Uncooperative behavior by us would lead to a fast reponse of uncooperative behavior of the partner.

I don't agree   1. ☐   2. ☐   3. ☐   4. ☐   5. ☐   I agree

b) Uncooperative behavior by us would lead to high costs for us due to image loss.

I don't agree   1. ☐   2. ☐   3. ☐   4. ☐   5. ☐   I agree

c) Uncooperative behaviour by us could lead to long-term damage to our company.

I don't agree   1. ☐   2. ☐   3. ☐   4. ☐   5. ☐   I agree

d) Uncooperative behaviour by the partner would lead to high damage to our company.

I don't agree   1. ☐   2. ☐   3. ☐   4. ☐   5. ☐   I agree

52. Please assess the following claims regarding your assessment of the **future relationship** with the partner:

A) The JV will remain stable for the whole time period specified in the JV contract.

I don't agree   1. ☐   2. ☐   3. ☐   4. ☐   5. ☐   I agree

b) After the JV contract ends, there will still be mutual dependency between us both.

I don't agree   1. ☐   2. ☐   3. ☐   4. ☐   5. ☐   I agree

53. Please assess the following claims regarding the **involvement** of your **top management**:

a) When negotiating the JV contract, our management had a higher bargaining power than the partner did.

I don't agree   1. ☐   2. ☐   3. ☐   4. ☐   5. ☐   I agree

b) How many times do senior executives from your firm and the partner firm typically meet per year?
_____ times per year.

7. INFORMATION REGARDING YOURSELF

Finally, we would like to ask you a few questions regarding yourself.

54. Where do you work?

☐ Headquarters   ☐ Joint Venture   Other: _____

55. Which position do you hold in your company?

☐ Board member   ☐ CEO   ☐ Director of a business division
☐ Area manager   ☐ Director of a department ☐ Group leader
☐ Assistant to the management   Other: _____

56. In which functional area do you work?

☐ Production   ☐ Marketing   ☐ Sales   ☐ Controlling
☐ General Management   Other: _____

That was the last question. Thank you very much for your having taken part!

Should you so wish, we can send you an **Executive Summary** *of our results. In that case, we need your postal address. The address data wil be treated confidentially and in no case will it be forwarded to third parties.*

Name: Company:

Position: Email:

Address: Postal code:

City: Country: